User Friendly: Hands-On Exhibits That Work

ISBN 0-944040-22-5

Association of Science-Technology Centers
1413 K Street, N.W., Tenth Floor
Washington, DC 20005-3405
Telephone: (202) 371-1171

User Friendly: Hands-On Exhibits That Work

Jeff Kennedy

Association of Science-Technology Centers
Washington, D.C.

About ASTC

The Association of Science-Technology Centers (ASTC) is a nonprofit organization of museums and related institutions dedicated to increasing public understanding and appreciation of science and technology.

ASTC and its members recognize the unique potential of science museums to provide informal science education through programs and interactive exhibits. ASTC seeks to improve the operations of science museums, to facilitate communication among its members, and to help museums broaden and diversify their audiences as well as serve as educational resources to their communities.

ASTC's major services include organizing and circulating hands-on science exhibitions to North American museums; publishing a bimonthly newsletter and other related literature; conducting an annual conference and workshops on various topics; representing science centers with Congress, government agencies, industry, foundations, and other organizations concerned with science education; conducting surveys according to member needs on such topics as salaries, science center operation costs, and programs; and administering a Reciprocal Free Admission Program, in which participating museums waive general admission fees for one another's members.

ASTC currently has nearly 350 members and represents some 30 countries. Although ASTC members vary in size, scope, and style, all are committed to increasing science literacy.

ASTC is supported by member dues, fees for services, and grants from federal agencies, private foundations, and corporations.

Designing hands-on exhibits is a complex process that, at best, draws on the insights and skills of many people. There are subject area experts to consult, educators and evaluators, writers and editors, graphic and industrial designers, carpenters and technicians of all kinds, and sometimes computer programmers and filmmakers as well. In the end, the measure of success is how well the exhibit works for the wide variety of people who visit the museum. It is easy and enjoyable to use? Is it clear what it's all about? Do people go away with a feeling that they understand something more than they did before they encountered the exhibit?

These questions are complex and difficult to answer, and, as author Jeff Kennedy notes, it is not often we feel we have succeeded completely. This publication should help you to make sure that one category of concerns at least is addressed: the multiple issues that, in the jargon of industrial design, are called "human factors."

Human factors – the study of the relationship between people and machines – has helped industrial designers to improve the effectiveness, reliability, safety, and ease of use of many machines and consumer products, from food processors to airline cockpits. Human factors has also led to improvements in products for older people and those with disabilities. In exhibits, sharp corners, confusing instructions, glare from reflective materials, and controls that are too high or too hard to turn are design problems that are all too common, but that might be avoided by greater attention to these concerns.

In commissioning this study, the Association of Science-Technology Centers asked designer Jeff Kennedy to look for ways that the discipline of human factors might be applied to the design of hands-on exhibits in museums. A review of existing literature and, more important, case studies at several museums formed the basis for developing these guidelines. We would like to thank Boston's Museum of Science, the Franklin Institute Science Museum, the New York Hall of Science, and the Ontario Science Centre in particular for their willingness to participate in this study in the interest of enhancing the quality of hands-on exhibits. To ensure that these guidelines would promote the goal of designing exhibits for *all* visitors, the author also relied on the advice of Betty Davidson of the Museum of Science for insight into special needs of visitors with disabilities. We would also like to thank Sheila Grinell, Kathleen McLean, Renee Wells, and Kathy Wilson for reviewing the manuscript and offering helpful suggestions.

Our final but foremost thanks go to the Institute of Museum Services, whose Professional Services program provided us with the support to make this publication possible.

Wendy Pollock

ASTC Deputy Director
Exhibits and Programs

About the author

Jeff Kennedy has been designing hands-on exhibits for 15 years. He received an undergraduate degree from Brown University and a Master of Industrial Design degree from the Rhode Island School of Design. His exhibit design career started at the Oregon Museum of Science and Industry (OMSI), where he was senior exhibit designer from 1975 to 1980. In 1980 he moved to Boston and from 1982 to 1984 worked at Cambridge Seven Associates, an architectural firm specializing in museum design. Jeff Kennedy Associates was formed in 1984. The firm focuses on the planning and design of museum exhibits interpreting scientific, technological, and historical subjects. Jeff has lectured on exhibit design at Massachusetts Institute of Technology, Harvard University, Boston University and Rhode Island School of Design and has spoken at several ASTC conferences.

It is embarrassing to create an exhibit that has a glaring design error, but we have all done it. After having a small, but obvious dimensional error pointed out to me, I decided to develop a human dimension checklist as a reference tool for our office. Over time the checklist expanded to include other "human factors" considerations and eventually, with the help of ASTC and the Institute of Museum Services, developed into this publication.

Human factors engineering was developed primarily in weapons research projects during World War II. More recently, significant contributions by human factors specialists have been made in aerospace projects, and someone from this field often plays a key role on teams that develop a new consumer product. The purpose of human factors, generally, is to study the relationship between people and equipment within the context of the environment in which the equipment is used. All factors affecting the human/equipment interface are considered, including physical and mental processes, efficiency, safety, and reliability.

The all-encompassing view of human factors engineering is perhaps what strikes a familiar chord with hands-on exhibit design. We too need to consider a user's physical and intellectual interaction equally. And we need to consider the relationship between exhibits and the environments in which they are used. Finally, there is this striking similarity: behavioral testing of the human user is fundamental to the human factors design process. Only by observing human behavior through the use of prototypes or mock-ups can designers of hands-on exhibits anticipate how a hands-on exhibit will be used by museum visitors.

In practice, the discipline of human factors is often highly technical and specialized. For the purposes of this publication, it is important to look at the big picture, seeing how the basic concerns and issues of this field can be applied to the special problems of hands-on exhibit design. Perhaps most importantly, human factors is a source of inspiration for placing the needs of the user at the center of the exhibit design process. By identifying and successfully resolving all of the small exhibit-user issues in each hands-on exhibit, I believe we can greatly improve visitors' use and understanding of hands-on exhibits, and make their time in the museum more enjoyable and rewarding.

Credits

Editor
Wendy Pollock

ASTC Production Coordinator
Susan McCormick

Human Factors Consultants
Bob Luchetti, Kurt Bodden
Robert Luchetti Asscociates

Design
Geoffry Fried

Production
Joan Louis

Research Assistance
Tom Kirby, Susan Shelkrot

Diagrams
Jeff Tate
Sumida Design, Boston

Printing
United Lithograph, Somerville

Photographs taken at the following locations:

AMERICA CREATES: 200 YEARS OF PATENTS AND COPYRIGHTS: Page 22 (top)

FRANKLIN INSTITUTE SCIENCE MUSEUM, PHILADELPHIA: Pages 11, 14 (left), 15, 17, 23 (bottom), 25 (right), 44, 54, 55, 62 (right)

MUSEUM OF SCIENCE, BOSTON: Pages 19 (left), 26 (bottom), 29 (left), 33, 49 (top), 67

NEW YORK HALL OF SCIENCE, CORONA: Pages 26 (top), 28, 30, 31 (left), 32, 42, 61 (right), 63

NORTH CAROLINA MUSEUM OF LIFE AND SCIENCE, DURHAM: Pages 18 (left), 20, 61 (left), 64

ONTARIO SCIENCE CENTRE, TORONTO: Pages 5, 9, 13, 18 (right), 21, 22 (bottom), 23 (top), 24, 25 (left), 29 (right), 36, 37, 38 (left), 39, 41, 49 (bottom), 50, 51, 52, 53, 60, 62 (left)

REUBEN H. FLEET SPACE THEATER AND SCIENCE CENTER, SAN DIEGO: Page 43

THE SCIENCE MUSEUM, LONDON: Page 31 (right)

Photograph on page 38 (right) courtesy of National Geographic Society

Photograph on page 68 courtesy of the California Museum of Science and Industry, Los Angeles

Contents

1

The Challenge of Hands-On Exhibits

At the conclusion of a recent hands-on exhibit brainstorming session, an experienced colleague remarked that it isn't success we should be aiming for, but to fail less miserably. This tongue-in-cheek comment is a painful reminder of what a formidable challenge it is to create a successful hands-on exhibit.

The people responsible for creating exhibits are often too close to their work to be objective. As professionals, it is easy to get sidetracked, creating exhibits for ourselves rather than for visitors. But the visitor is ultimately the judge of how well we have done our work. It is critical that we become and remain students of human behavior, incorporating what we learn into the exhibits we create. Here are some observations which I have kept in mind in developing these guidelines about museum visitors.

• Museum visitors are a diverse lot, including people of all ages, sizes, and physical and intellectual capabilities, as well as varied interests and cultural backgrounds. Exhibits should be designed to accommodate and appeal to as wide a range of visitors as possible.

• Visitors find ways to interact with exhibits which the designer never anticipated. This is an advantage for exhibits that are intended to be open-ended. But exhibits that are not usable or not understandable, when operated in ways not originally intended by the designer, can lead to frustration or miscommunication. The blame in this situation lies with the designer, not the user. Hands-on exhibits should never frustrate, inhibit, confuse, or frighten visitors.

• Designers of hands-on exhibits need to become students of visitor behavior. There are several ways to do this:

Spend time on the museum floor observing people interacting with exhibits. Visit a variety of museums and see how visitor behavior varies.

Incorporate front-end evaluation into the exhibit design process. Find out what people are interested in and what they already know (or don't know).

Build prototypes and write prototype labels. Evaluate their use by visitors. Integrate results into the design.

Evaluate visitor interaction with the finished exhibit. Make modifications if needed. Apply results of your evaluation to your next project.

WHAT IS A "HANDS-ON" EXHIBIT?

The term "hands-on" in this publication refers to exhibits that require some kind of physical interaction between the visitor and the exhibit. This includes simple button pushing in computer and interactive video exhibits as well as more complex, multidimensional activities. "Hands-on" is different from "interactive" or "participatory," neither of which necessarily implies physical contact with the exhibit. These two terms readily accommodate the idea that exhibits can engage visitors in deep mental interaction without touching — an idea that this publication endorses. The focus here, however, is on human factors affecting the design of exhibits in which interaction is physical as well as intellectual.

Every museum has its own unique setting which reflects its mission, philosophy, staff expertise, and resources. Beyond this, there are a number of characteristics of the museum setting that have direct bearing on the design of hands-on exhibits.

• Museum visitation varies greatly both seasonally and day to day. Exhibits must be designed to serve visitors effectively even during the museum's busiest days – providing ample room for operating and for moving comfortably among exhibits; providing acoustical controls which prevent an exhibit area from becoming unbearably noisy when crowded with enthusiastic exhibit users.

• Exhibits sometimes begin with an introductory film, or end with a concluding experience. Other than this, however, visitors in hands-on museums typically select their own path of circulation. As a result, there is seldom a predictable, linear route through an exhibit. This means that each component must be designed to be usable and understandable on its own, even though it may be conceptually linked with other components.

• Sooner or later, most visitors suffer from "museum fatigue." Fatigue can be kept to a minimum by carefully orchestrating the exhibit experience to provide changes in pace, intensity, and appearance. Frequent rest stops also need to be designed into the visitor experience.

• In larger museums, there may be so many exhibits that they compete with each other for limited visitor time. Aesthetics is one factor often associated with the attracting power of an exhibit. Exhibits which are recognized as being usable and understandable also attract visitors.

EVERYONE IS A DESIGNER

Good hands-on exhibits are the result of a team effort which can include administrators, program developers, evaluators, content specialists, curators, educators, interactive video and computer experts, exhibit designers, artists, and fabricators. Throughout this publication, the terms "design" and "designer" refer to the activity of all members of the team responsible for creating an exhibit, as well as to the specific tasks of the "exhibit designer."

Challenge three: knowing the medium of hands-on exhibits

Some hands-on exhibits are open-ended, inviting experimentation and exploration, while others interpret a single phenomenon in a controlled way with the goal of eliciting an "ah-ha" reaction. Some are serious, and others are lighthearted. The variety of experiences afforded by hands-on exhibits makes it possible to reach a diverse audience.

• Hands-on exhibits can involve all of a visitor's senses: vision, smell, hearing, touch, taste. This array of sensory experiences also allows utilization by people with physical and sensory impairments.

• How to use hands-on exhibits should be evident to users through nonverbal cues. The design must provide users with a good mental model of the exhibit's operation, make clear the relationship between controls and their effects, and provide feedback that makes the results of actions clear and obvious.[1]

• Not all concepts lend themselves to interpretation through hands-on exhibits. When the hands-on approach is forced upon a concept, or vice versa, the exhibit is doomed to failure. One of the essential tasks of prototyping is to determine whether the exhibit conveys the intended concept.

• It is unrealistic to think that the first-generation design of any hands-on device will be ideal. It often takes five or six iterations to develop a consumer or industrial product with optimum user and functional characteristics. One of the challenges inherent in hands-on exhibit design is that very often the end product is one-of-a-kind. The best way to optimize first-generation exhibit performance is to integrate prototyping and evaluation into the design process.

• Being able to talk with interpretive staff can enhance a visitor's understanding of exhibits. Interpretive staff can also expand a museum's expertise in exhibit design by observing how visitors interact with exhibits, and by contributing their insights during the exhibit design process. Unfortunately, not every exhibit can be staffed, and hands-on exhibits typically must be designed to be totally self-sufficient.

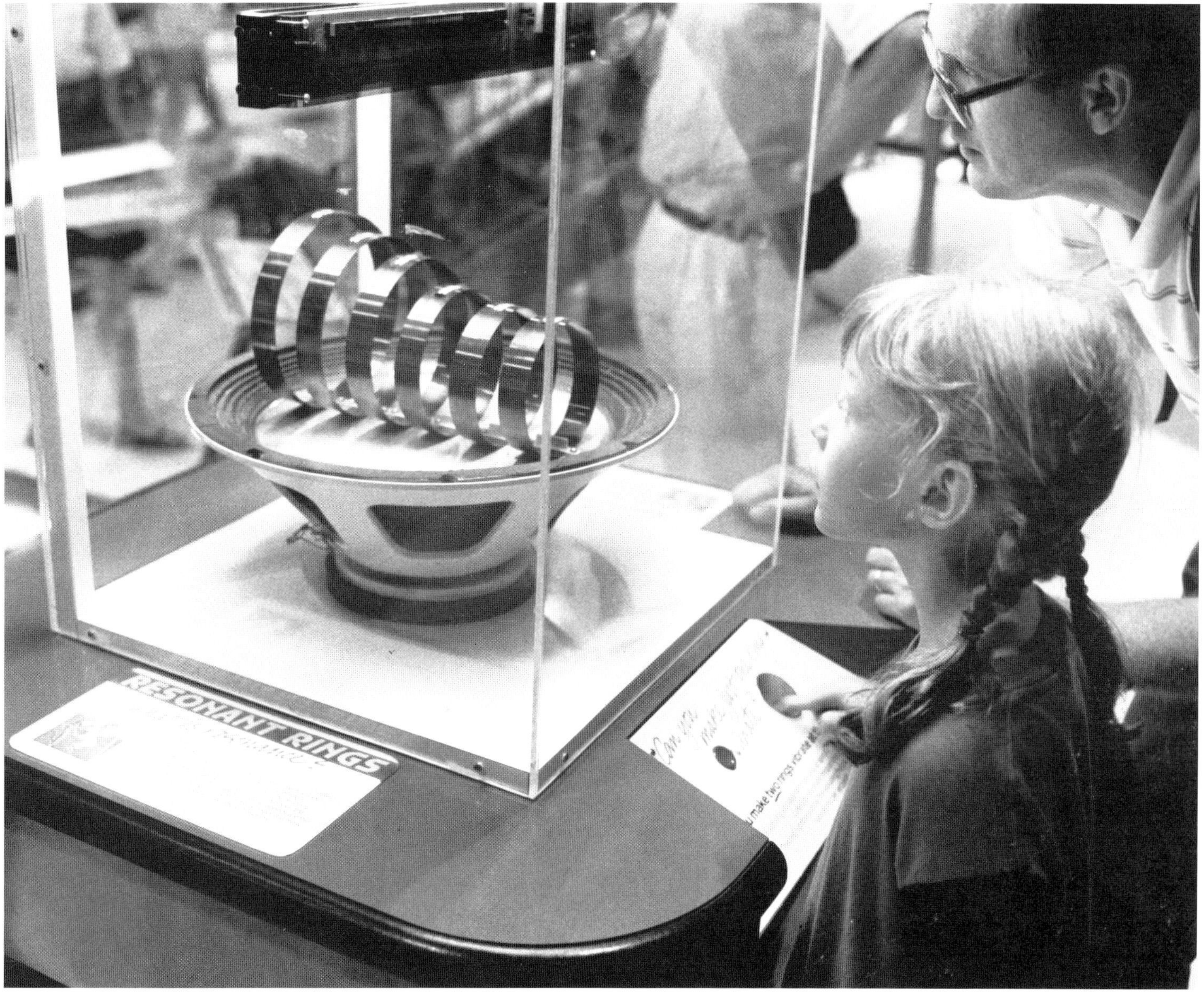
RESONANT RINGS

2

Human Measurement: Fitting the User

Although size seems like a straightforward matter, it is surprising how often the dimensional aspect of an exhibit is clearly wrong. Getting the size of the exhibit and the physical placement of components dimensionally correct is fundamental to user comfort.

Anthropometric guidelines for common exhibit configurations

One of the problems with using a generalized anthropometric data base as a reference source is that very little of the data is relevant to the design problems at hand. As a result, large reference books remain on the shelf most of the time. The anthropometric information which follows has been selected for its direct relevance to hands-on exhibit design.

The following dimensional guidelines apply to exhibits that need to accommodate a diverse audience including children, adults, and wheelchair users. For most hands-on exhibits, it is possible to build a single unit which will accommodate this range adequately, if not altogether ideally. For certain interactive tasks, however, a fixed size exhibit may not be an acceptable solution. Some alternatives to a fixed size exhibit are discussed in the next section.

The following configurations are widely used in the presentation of hands-on exhibits. Each diagram shows use by a five year old male, a twelve year old female, an adult female wheelchair user, and an adult male.

A range of sizes Exhibits must be designed to fit a wide range of visitor sizes. The great variation in size between a five year old and an average adult male is evident in this diagram. Variation in height presents the greatest problems for exhibits designed to be operated while standing.

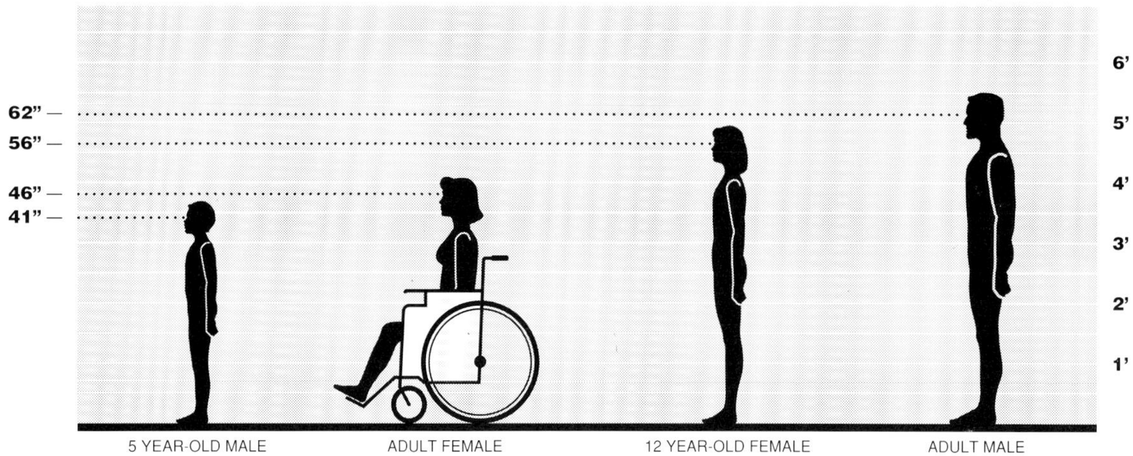

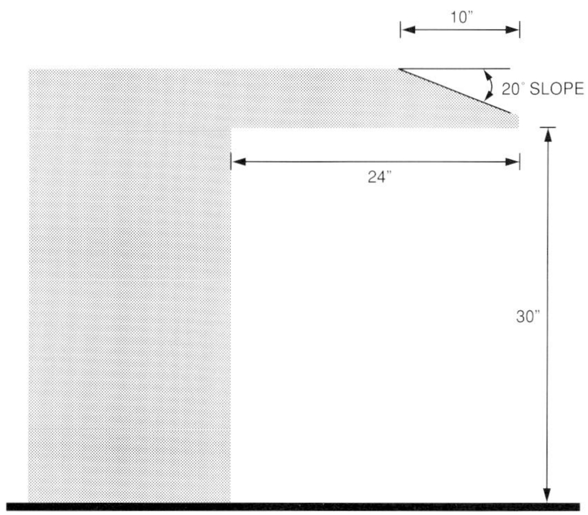

Table standards The table dimensions shown are standard for all tables in the following diagrams. The 30-inch dimension from the floor to the underside of the table is based on clearance required for a standard wheelchair arm. The 10-inch depth, or forward-reach dimension, allows small children and people in wheelchairs to reach exhibit controls. The 20-degree angled control panel allows easy reading of controls and labels by everyone.

Sitting at table or counter exhibits (top)
Seating is one of the best ways to deal with problems created by height disparity. The seat shown here is 18 inches high – standard in many museums. As the chart indicates, all users are able to use the tabletop exhibit when seated. A five year old would be the least comfortable. Exhibits that are popular with this younger age group are often designed to be lower.

Standing at tabletop or counter exhibits (bottom)
The standard table dimensions used for exhibits operated while sitting also work, in most cases, for exhibits operated while standing. This is important because it means that all exhibits can be accessible to wheelchair users. This diagram shows that all four users can comfortably reach the front control panel.

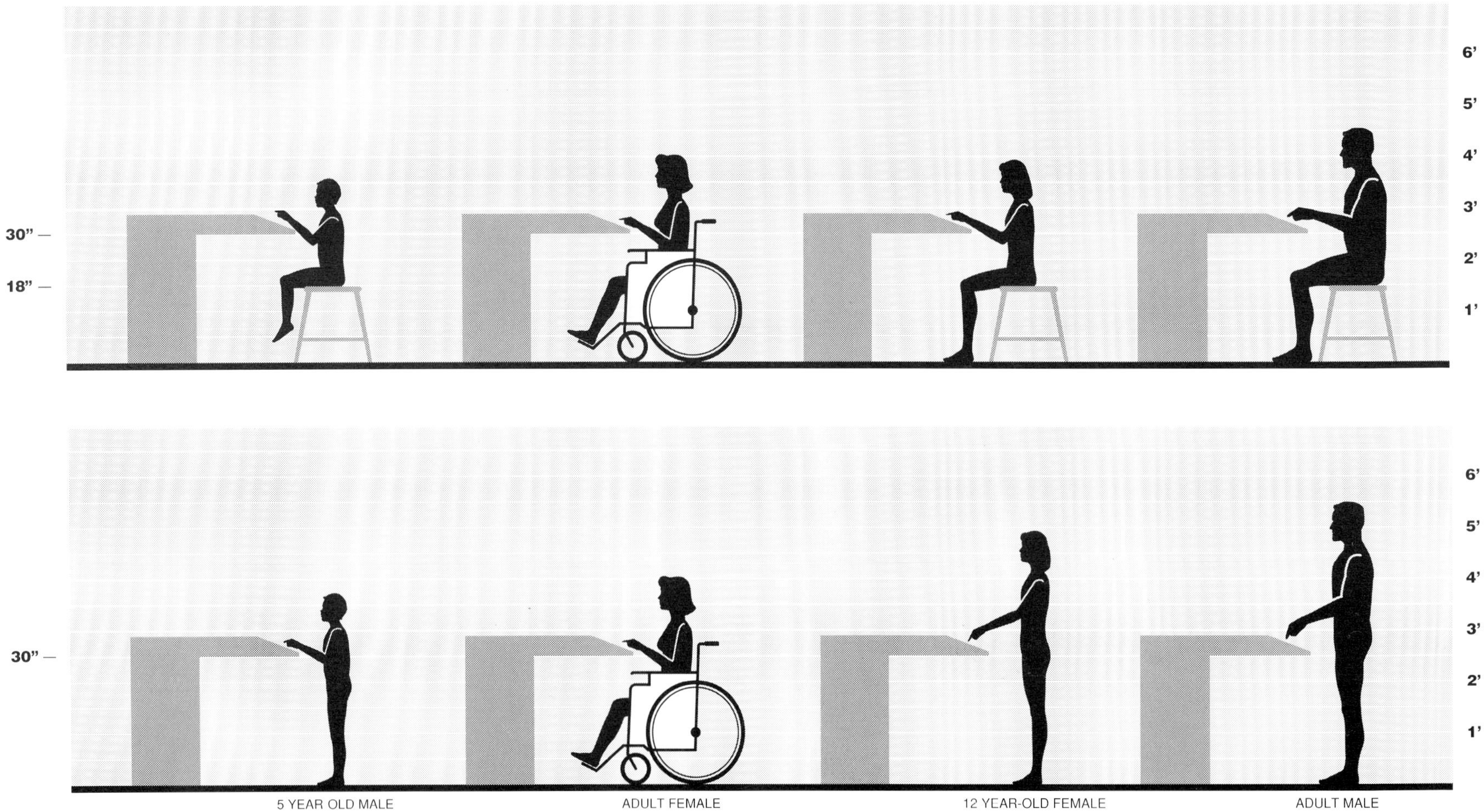

Feet off the floor
Data and charts can help analyze some human dimension design problems, but these tools have their limitations. For instance, only by observation would you know that five year olds can readily use most seated exhibits, even though their feet don't reach the ground.

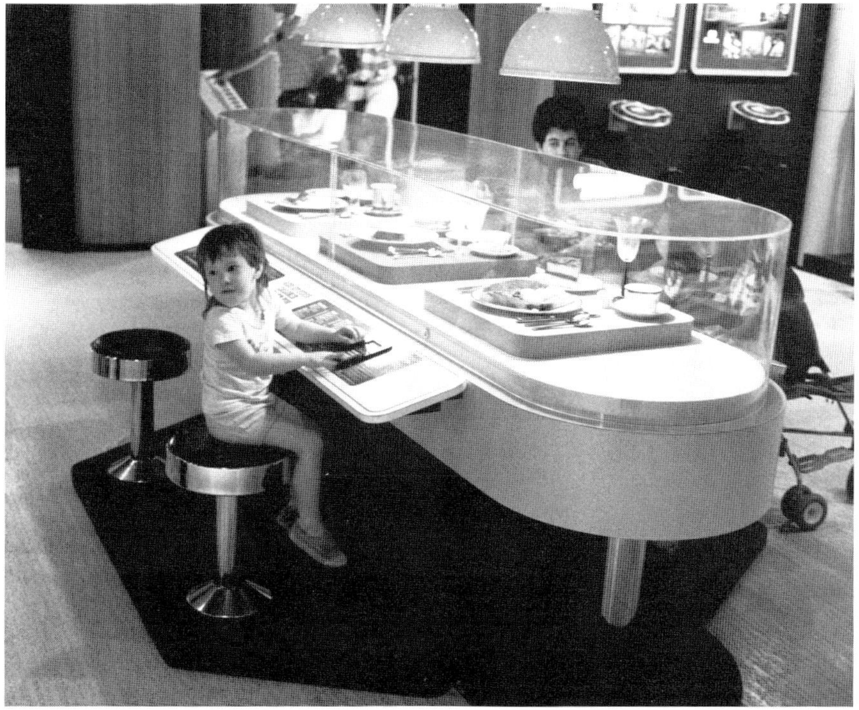

The kneeling position
Here is a classic example of an adult kneeling to look into a viewer mounted at tabletop height.
This problem can be addressed by providing seating, or having multiple units mounted at different heights.

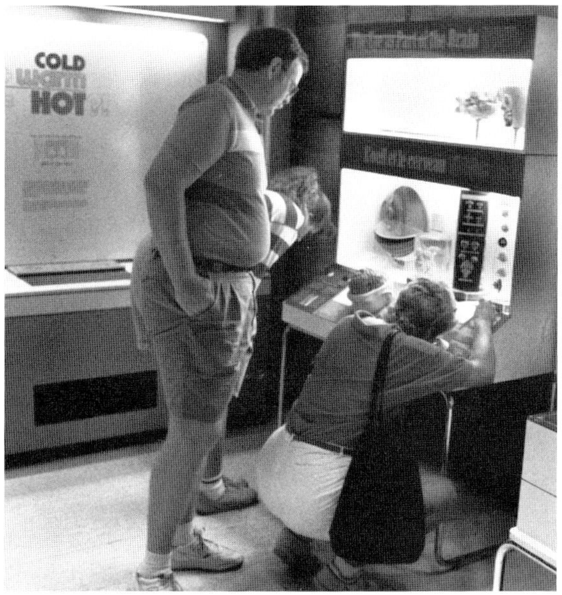

HUMAN MEASUREMENT DATA BASES

Physical anthropometry is a field of study developed in response to the military's need to have a data base of human measurements. This data base has been used for making a wide range of decisions, from determining standard uniform sizes to setting height and weight cut-offs for draftees. Anthropometric data bases are also used by a variety of professions including industrial designers, engineers, and architects. Nike Inc., a prominent manufacturer of athletic footwear, has compiled the world's most extensive data base on dimensions of the human foot. A good source of anthropometric data for exhibit designers is *Humanscale*, by Henry Dreyfuss Associates.[2]

Standing reach (top) Wall mounted exhibits need to consider the range of users' reach. This diagram shows that the optimal average reach zone for accommodating all users is between 36 inches and 48 inches above the floor.

Critical viewing at standing exhibits (bottom) When visitors stand close enough to an exhibit to operate controls, the optimal placement of critical visual material, such as labels, is in a relatively narrow zone, from 42 to 49 inches. This zone establishes centerline heights, not the overall size of the material being presented.

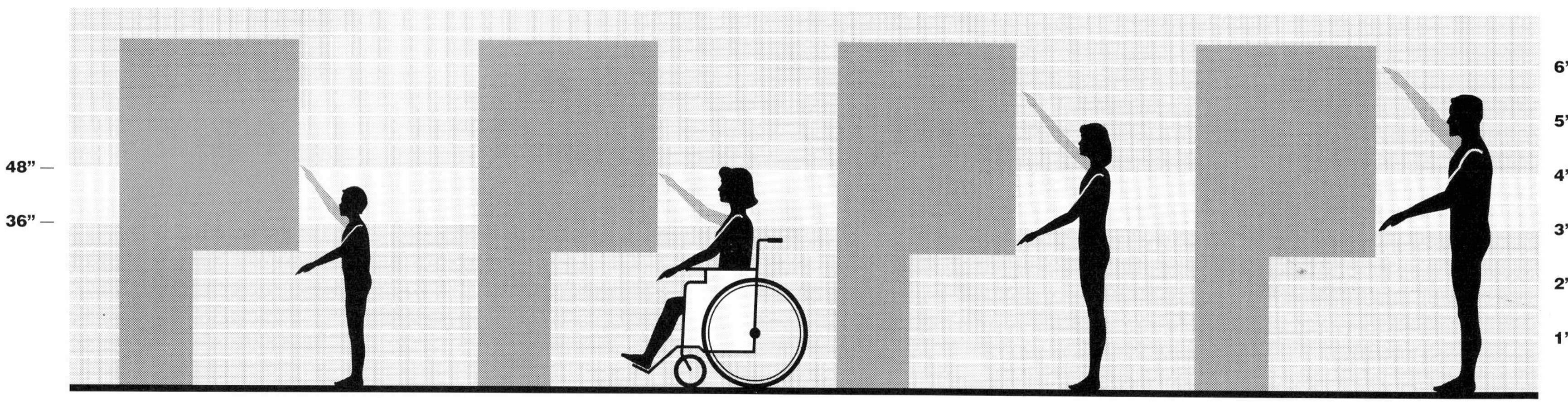

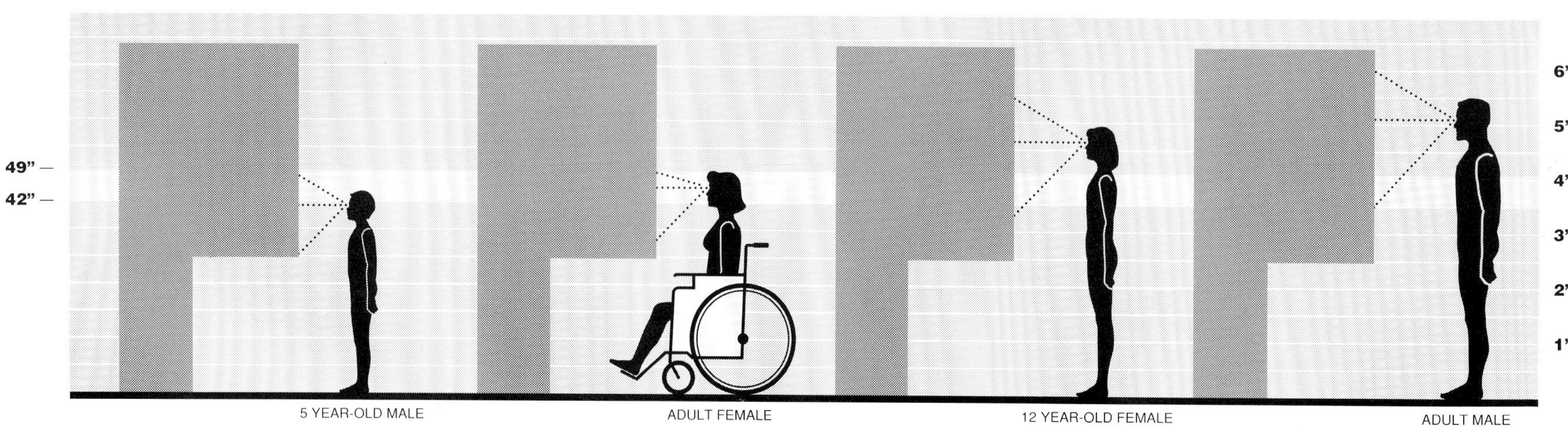

Vertical reach zone The exhibit on the right is in a comfortable reach zone. The one on the left, however, is mounted approximately eight inches higher and requires a strained reach.

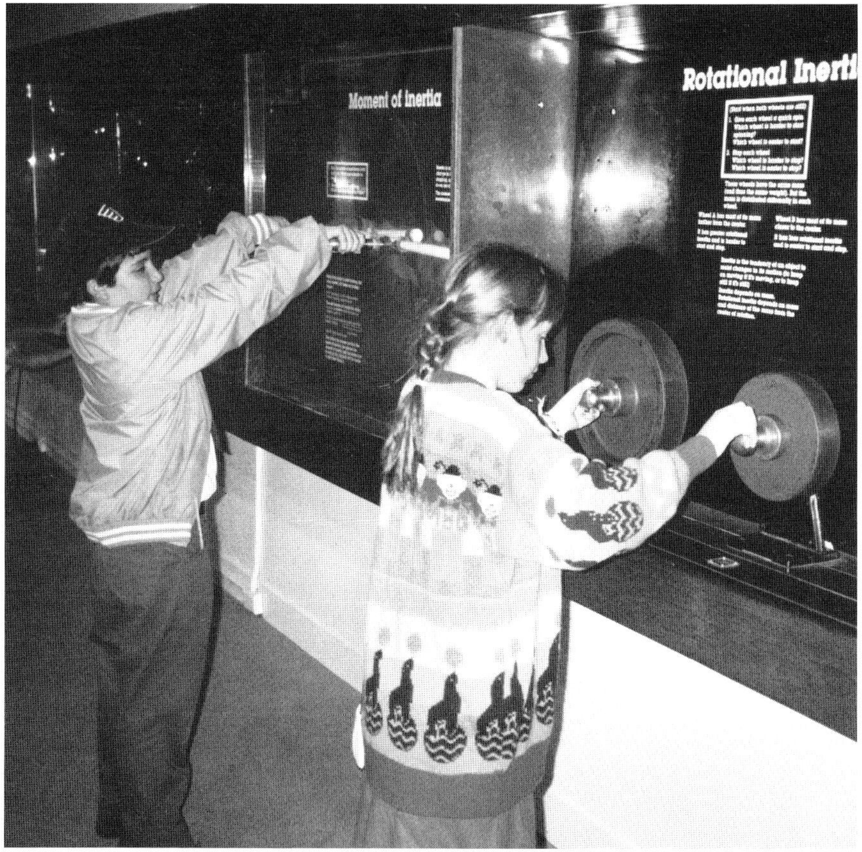

DESIGN GUIDELINES FOR PEOPLE WITH MOBILITY IMPAIRMENTS

The following guidelines were developed in response to the significant number of museum visitors who have either a short-term or long-term mobility impairment including individuals using wheelchairs, canes, and walkers. These guidelines do not conflict with general design standards. In fact, items 4 and 5 suggest a standard which would benefit all visitors.

1. Wheelchair access to exhibits should always be from the front rather than from the side.
2. The minimum dimensions for an opening providing frontal wheelchair access to an exhibit are: height 30 inches, width 30 inches, and depth 28 inches. In addition, a minimum of 48 inches should be provided between the front of the exhibit and adjacent exhibits, to provide comfortable room for maneuvering a wheelchair into position.
3. Exhibit seating should not be permanent. Wheelchair users need to be able to move seating out of the way easily.
4. A sloped-top exhibit table makes it easier to see things from a seated position.
5. Controls located within 8 to 10 inches of the front edge will enable many users with restricted reach to use the exhibit.
6. Blind people who use canes navigate based on cues at floor level. Therefore, no part of the exhibit cabinetry above waist level should extend beyond the footprint defined by the perimeter of the exhibit base. This is a good, common-sense rule to prevent head bumping for all visitors.

Being able to accommodate users of different heights is one of the most common exhibit design problems. Sometimes a "one size fits all" approach is not feasible and an alternative design solution must be found. Several common solutions and their pros and cons are exaimined here.

• *Step-ups:* A small step, usually less than six inches high, is placed on the floor at the base of the exhibit. This is a low cost, low maintenance solution for creating an exhibit accessible to a range of sizes. The down side of step-ups is that they can get lost and are potentially a source of accidents if visitors fall off or trip over them.

• *Multiple units at varied, fixed heights:* Part or all of the exhibit is replicated at multiple heights, typically three to four, depending on the type of interaction required. This approach accommodates a wider range of users than step-ups and is somewhat more expensive.

• *Mechanically adjustable exhibit:* The exhibit or a part of the exhibit can be mechanically raised or lowered. Sometimes this is achieved by tilting the exhibit on a pivot point. This approach can accommodate a wide range of users. Depending on the type of mechanical adjustment provided, cost and maintenance can be significant.

• *Seating used as a size equalizer:* Many exhibits require looking into viewers, or placing the head in a specific location to get a desired point of view. One of the best approaches for equalizing user height is to use a tabletop configuration and let visitors control two variables: the distance from the seat to the table, and the angle of their backs. This is done intuitively, and requires no instructions. This approach is recommended, however, only for exhibits where the duration of interaction is relatively short.

• *Mechanically adjustable seating height:* Adjustable seating is can be obtained by using gas cylinder technology – as in office chairs – or screw mechanisms. The drawback is that the adjustability becomes one more thing that the user has to take care of. If every seated exhibit had adjustable height control, users would have to make an adjustment every time they sat down. Given the short duration of most interactive tasks, adjustable height seating seems to be more bother than it is worth. Adjustable seating should, however, be considered for interactive video and computer exhibits where visitors spend substantial periods of time.

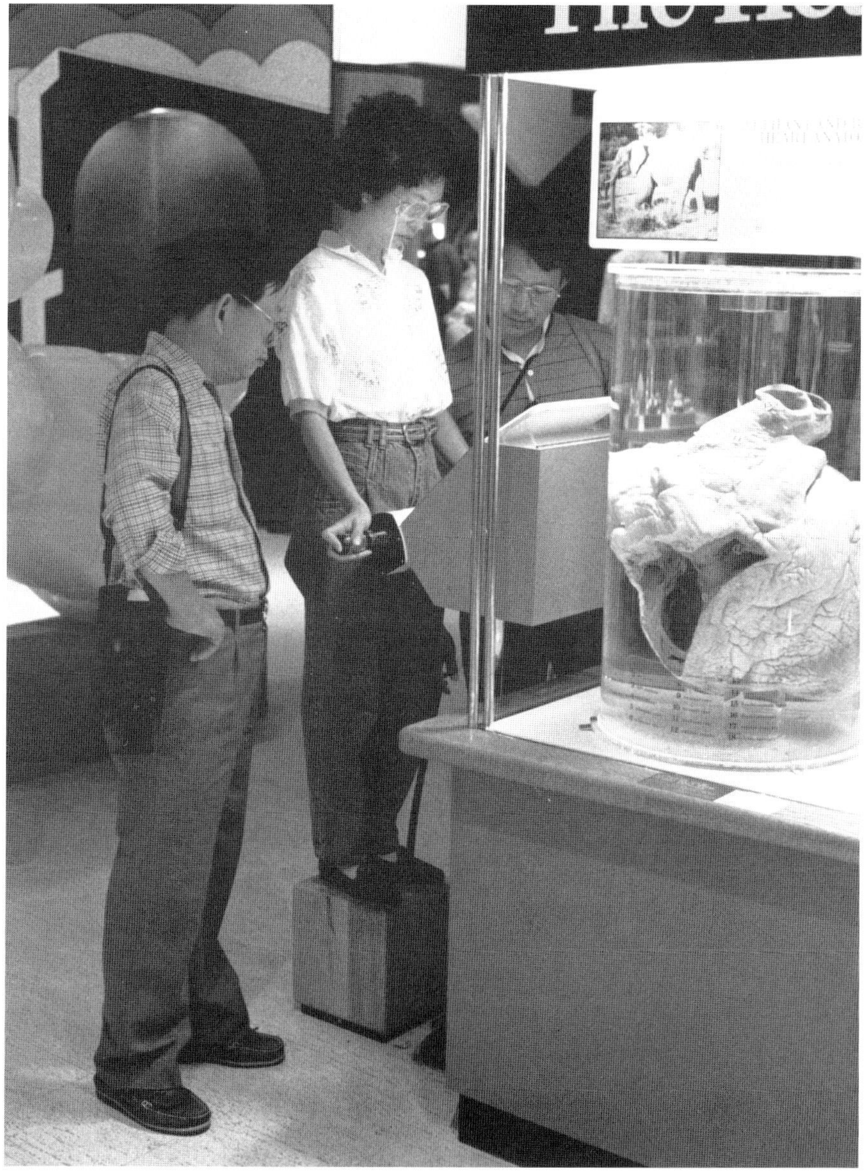

Multiple viewing heights This exhibit provides work stations at four different heights. The user on the right found a suitable height while the user on the left is at an uncomfortably low height. If designed for seated operation, this exhibit would have been more comfortable for a range of user sizes.

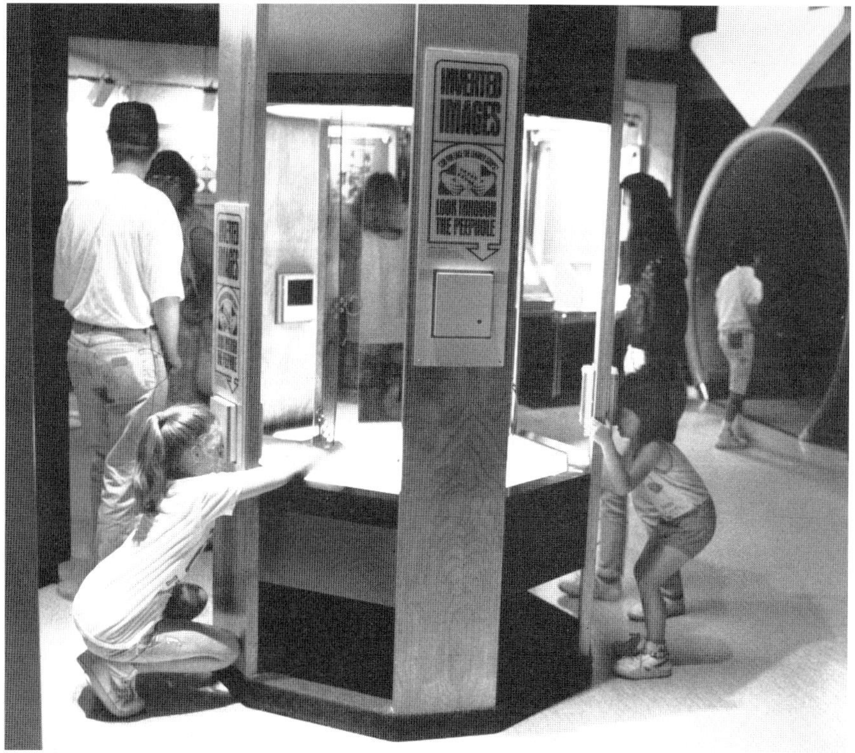

Step-up in action Step-ups allow small users to have access to exhibits. In this example, however, an adult stands on the step-up thinking it is necessary to operate the exhibit. In addition to creating this misconception, the step-up poses the problem of being a barrier to wheelchair users.

Tilt-top The top of this exhibit pivots, providing a good range of viewing angles. Unfortunately, the base below is a barrier to frontal access by wheelchair users.

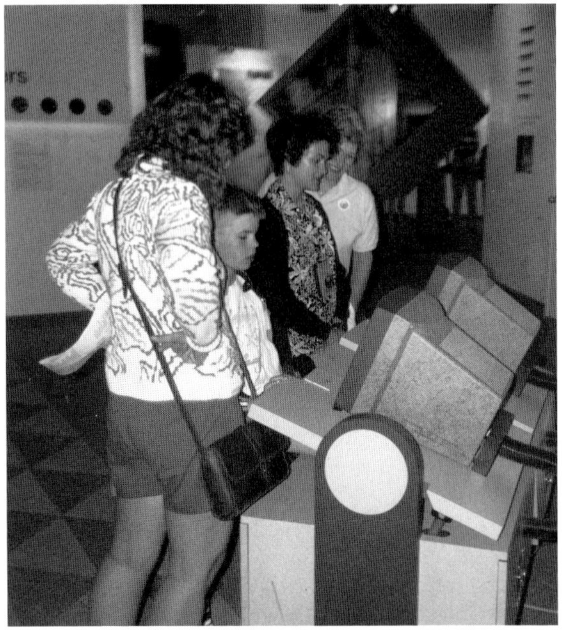

WHEN IN DOUBT, MOCK IT UP

Anthropometric data provide guidelines. In the actual practice of designing and making exhibits, it is essential to look at the specific requirements of the exhibit interaction and to consider all dimensions within this context. Whenever there is doubt about dimensions, a mock-up should be constructed. It is almost always worthwhile to make a full-size mock-up rather than a scale model. This allows you to get the actual feel of the exhibit.

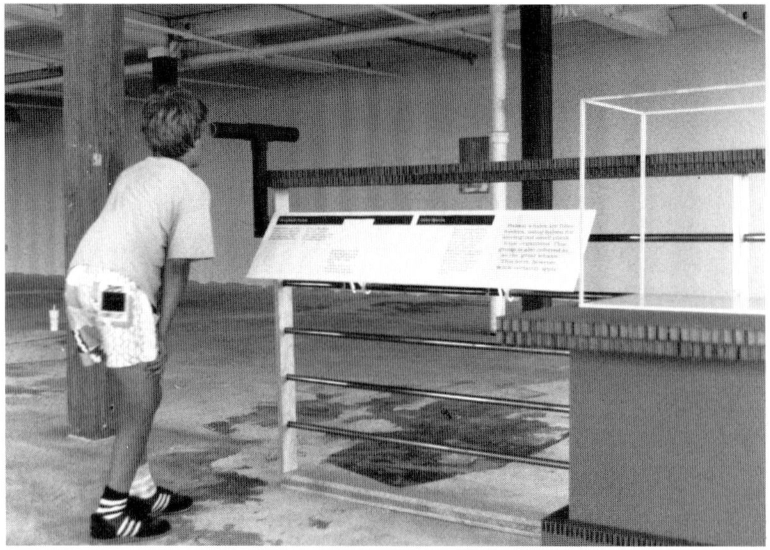

Full-size mock-up This full-size mock-up duplicated a section of permanent building rail which was a "given" in the exhibit space. The height and placement of graphics, artifacts, and hands-on exhibits were determined using the mock-up.

Tall user To adjust his height to the exhibit, this tall user has moved the stool back from the table and bent over at the waist. This position would be uncomfortable after a while, but is fine for the brief period of time spent here.

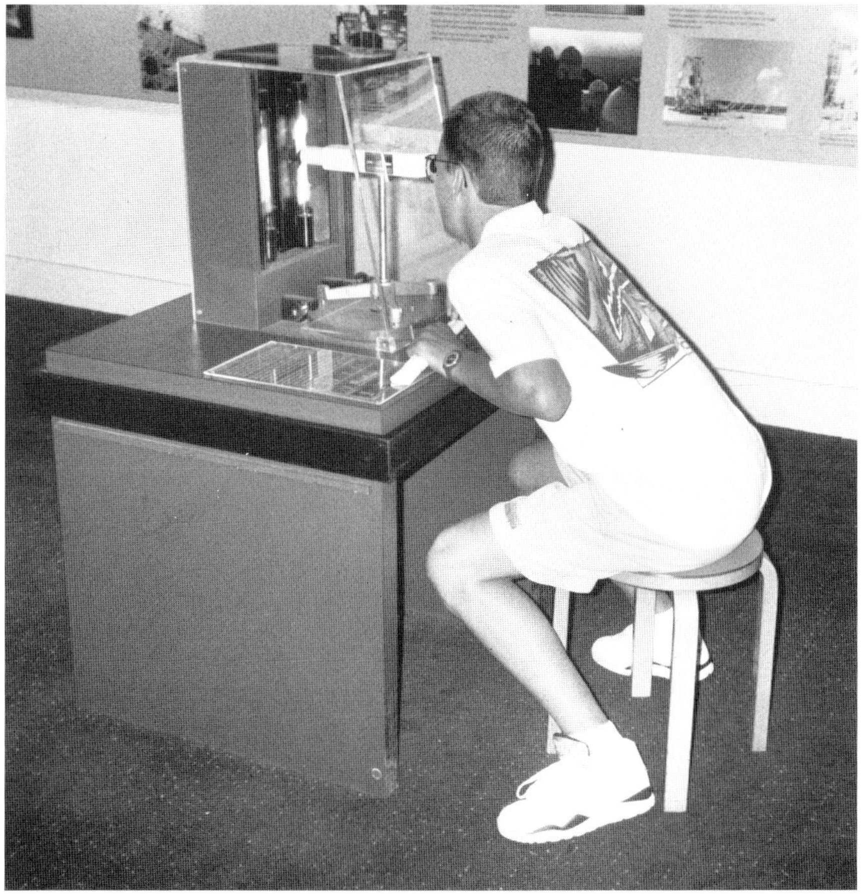

Short user This short user is sitting on the front edge of the stool to get closer to the exhibit. Little bending at the waist is required to be at a good viewing height.

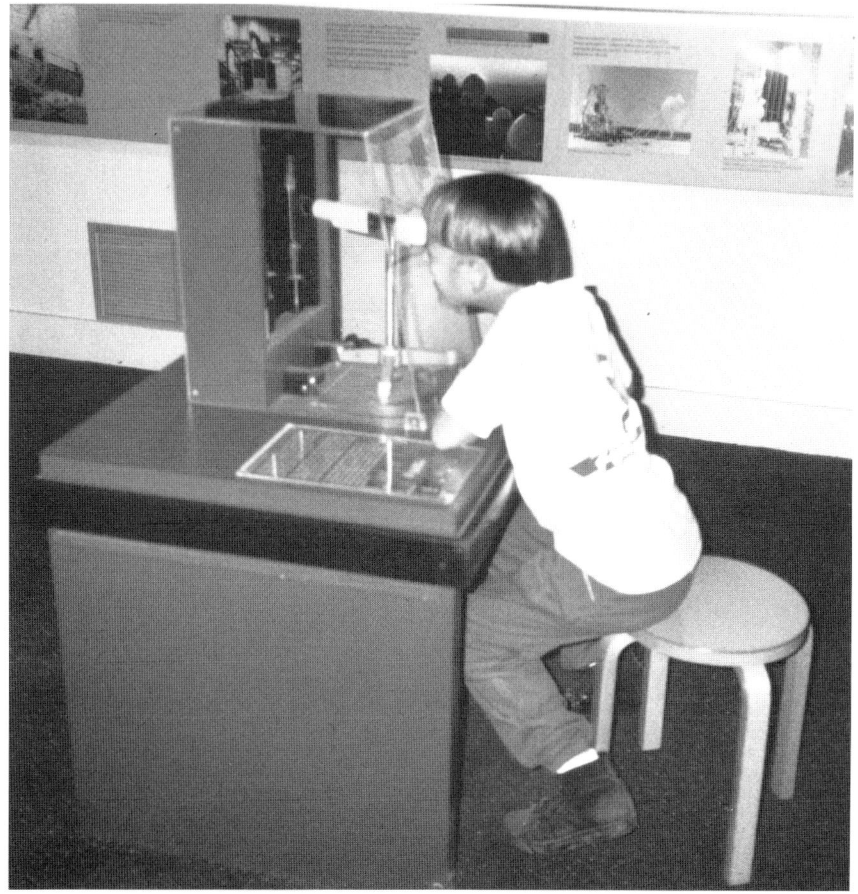

Controls

3

A visitor's ability to operate and understand a hands-on exhibit is closely linked with the design of the controls. It is important to keep in mind the diversity of people using hands-on exhibits, making sure that people of all ages and abilities have the opportunity to use the controls.

Although there are many interrelated factors to consider when designing controls, two issues are paramount: providing visual clues in control systems *and* choosing the right control for the job.

Providing visual clues in control systems

Combinatory poem This exhibit interprets the mathematical concept of permutations and combinations using a mechanical analog. Users turn a handle and watch as words are moved on a series of shafts, belts, and pulleys. All moving parts are exposed. The design approach optimizes the user's ability to make connections among exhibit control, operation, and concept.

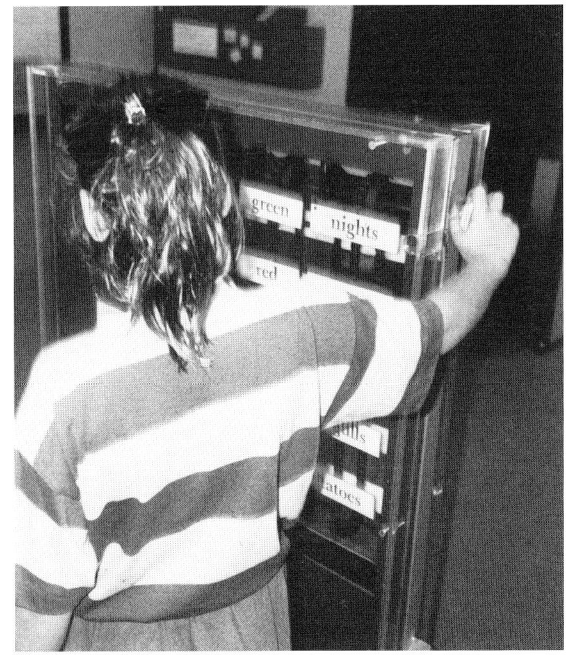

Visitors usually approach a hands-on exhibit by pulling, pushing, or turning whatever looks like a control device. This is followed by observing the resulting action, and finally, trying to figure out what it all means. Written instructions are often consulted only after a first pass at operating the exhibit in this intuitive way. It is clear that the success a visitor has using and understanding a hands-on exhibit is greatly influenced by visual clues provided by the exhibit's design.

Visual clues provided by the exhibit control system are particularly important since controls are the part of the exhibit with which visitors interact most directly. The following examples illustrate a variety of ways that visual clues can be incorporated into the design of exhibits.

Visible control systems

Making all of an exhibit's operating parts and control mechanisms clearly visible enables visitors to formulate a mental model of its operation. Exposing the linkages between controls and moving parts makes it easier for a user to see what control actions are possible, and to consider the likely results of these actions. Following are some examples of visible control systems.

Ship gears and propeller The connection between the control (the large round handle) and the exhibit's functioning parts (the gears and propeller) has been made very clear by making all parts visible. Visitors seem surprised and delighted when they find that turning the handle moves the giant propeller.

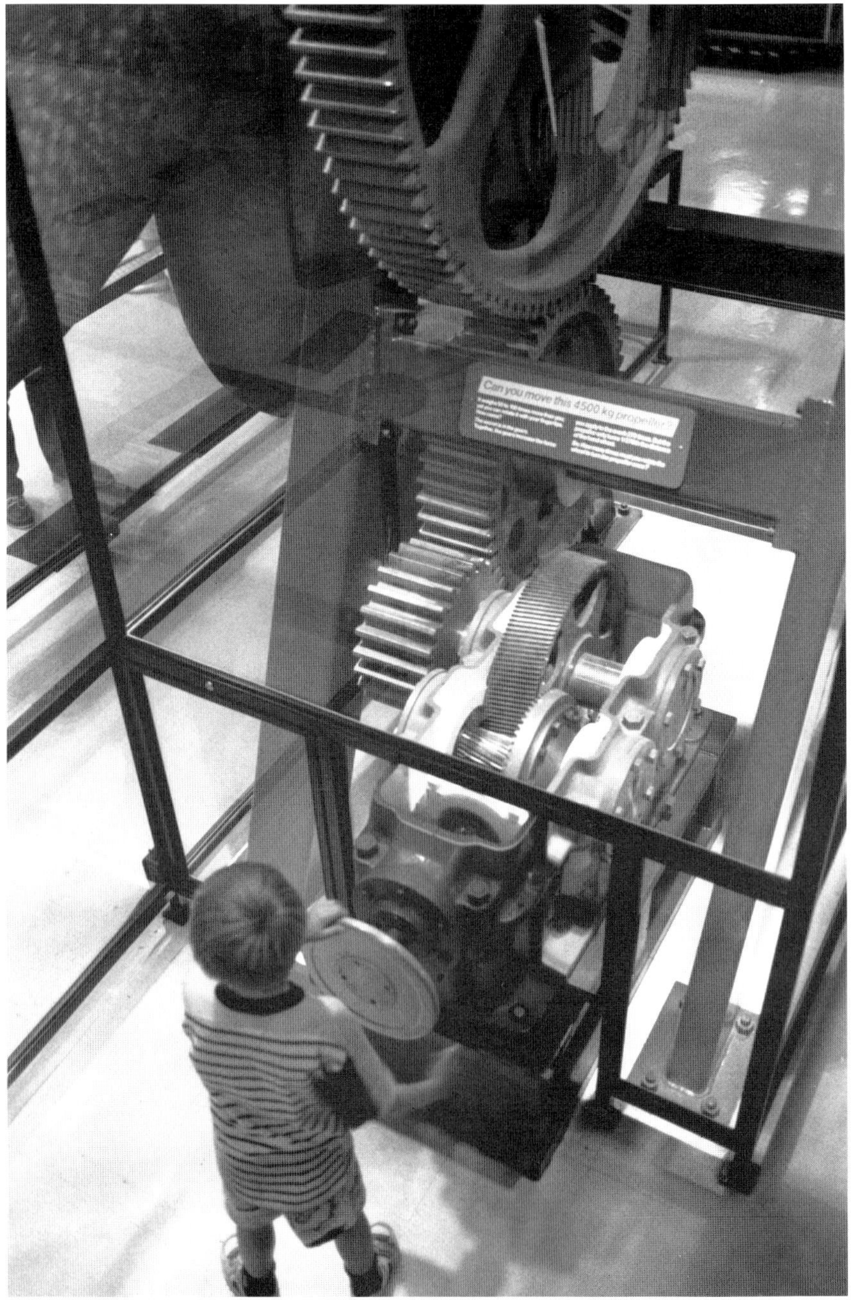

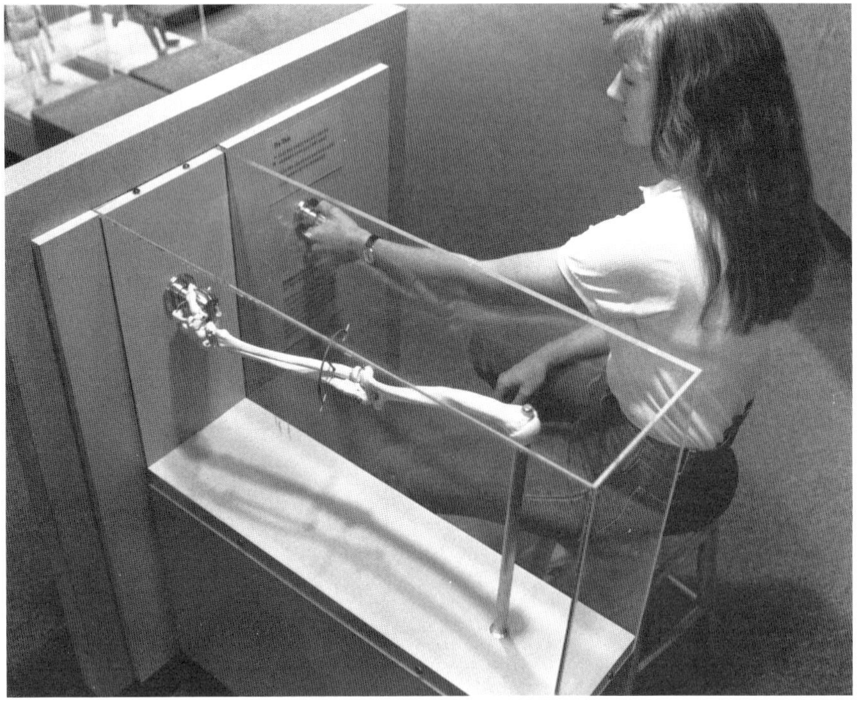

Inside your arm Here, the visual connection between the control device (the door knob) and the functional part of the exhibit (the skeleton arm) is apparent. Visitors readily anticipate that turning the door knob will make the skeleton move, and they delight in seeing their conceptual model borne out in practice. The mechanical linkage between the two door knobs is exposed on the opposite side, revealing how the user's arm causes the skeleton to move.

Reinforcing concepts with controls

Control design can sometimes provide visual clues that reinforce a key exhibit concept. This is demonstrated in the following two examples.

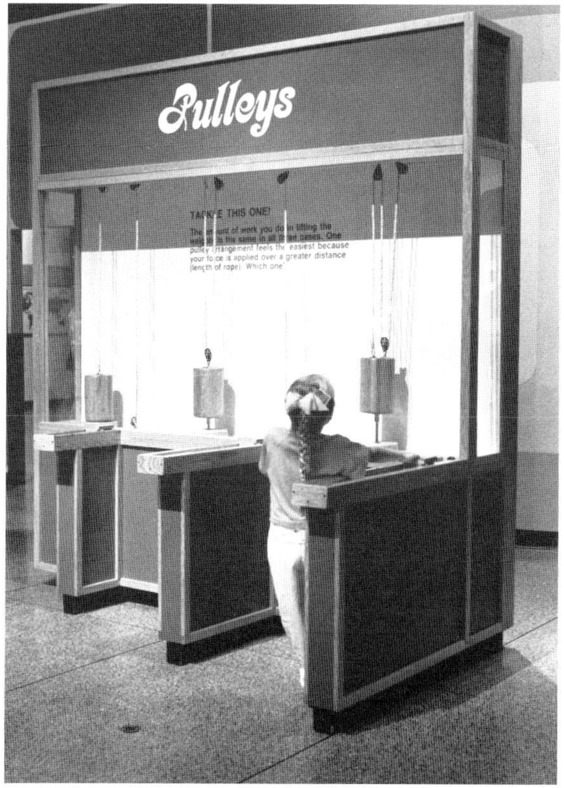

Pulleys In this exhibit, visitors find that the longer the rope is (and the more pulleys it goes through), the easier it is to lift the weight. The ropes are controlled by pulling wooden handles along tracks. An emphasis on the difference in the lengths of the three tracks provides a strong tie-in between the control and the scientific concept.

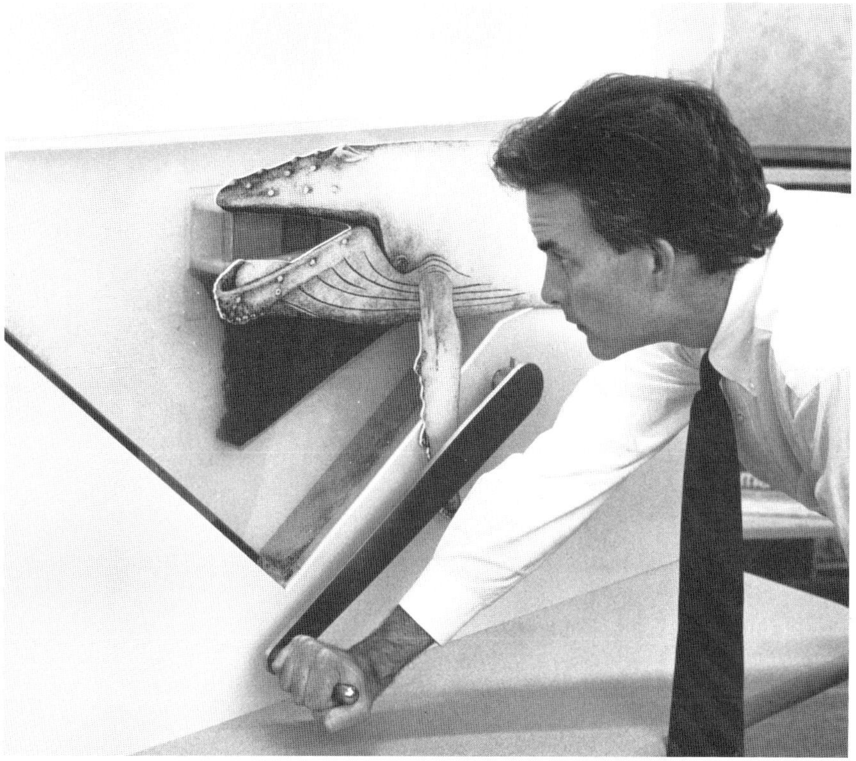

Humpback feeding
During feeding, a humpback whale's pleated throat expands tremendously as it takes in great quantities of water and food. In order to focus user attention on the expansion and contraction of the throat pleats in the exhibit, the movement of the control – the long lever – has been designed to exactly parallel the movement of the throat.

GUIDELINES FOR MAKING EXHIBIT CONTROLS UNDERSTANDABLE AND EASY TO USE

1. Minimize the number of controls. Fewer controls make exhibits look easier to operate and, in practice, they are easier.
2. When multiple controls are required, match the number of controls to the number of functions, and organize the controls according to function.
3. When multiple controls are required, make controls for different functions look different.
4. Make the action of the control relate as clearly as possible to the function being controlled.
5. Incorporate feedback into the control mechanism.

Use of the familiar

If all the exhibits in a museum had unique controls, the task of figuring them out would be exhausting. Fortunately, most hands-on exhibits can be operated using controls that are familiar to almost everyone. For example, round handles are for turning, buttons are for pushing, levers are for pulling. Use of familiar controls is a way of tapping into a visitor's existing knowledge, and helps to make the unfamiliar – the hands-on exhibit – familiar.

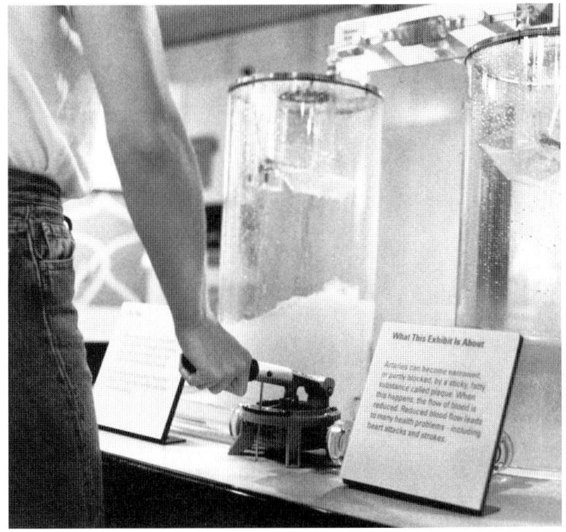

Getting a grip The lever on this hand-operated water pump is easily recognizable as a control for pulling and pushing. The bicycle grip helps identify it as something you are supposed to grasp.

Choosing the right control for the job

Controls are the link between a visitor and a hands-on exhibit. By choosing the right control for the job, designers can optimize a person's ability to use and understand an exhibit. Often, it is possible to make this choice only after a fully functioning exhibit prototype has been developed. It is critical to evaluate how well people are able to use controls and instructional labels during prototyping.

The following examples point out attributes of several types of controls that our case studies show to be in common use. The examples are divided into two categories: *mechanical* and *electrical.*

Mechanical controls

Mechanical controls transmit a user's energy to an exhibit. Included in this category are wheels, handles, and levers. In order to be accessible to all visitors, mechanical controls must require relatively little strength for their operation. User power requirements need to be evaluated during exhibit prototyping.

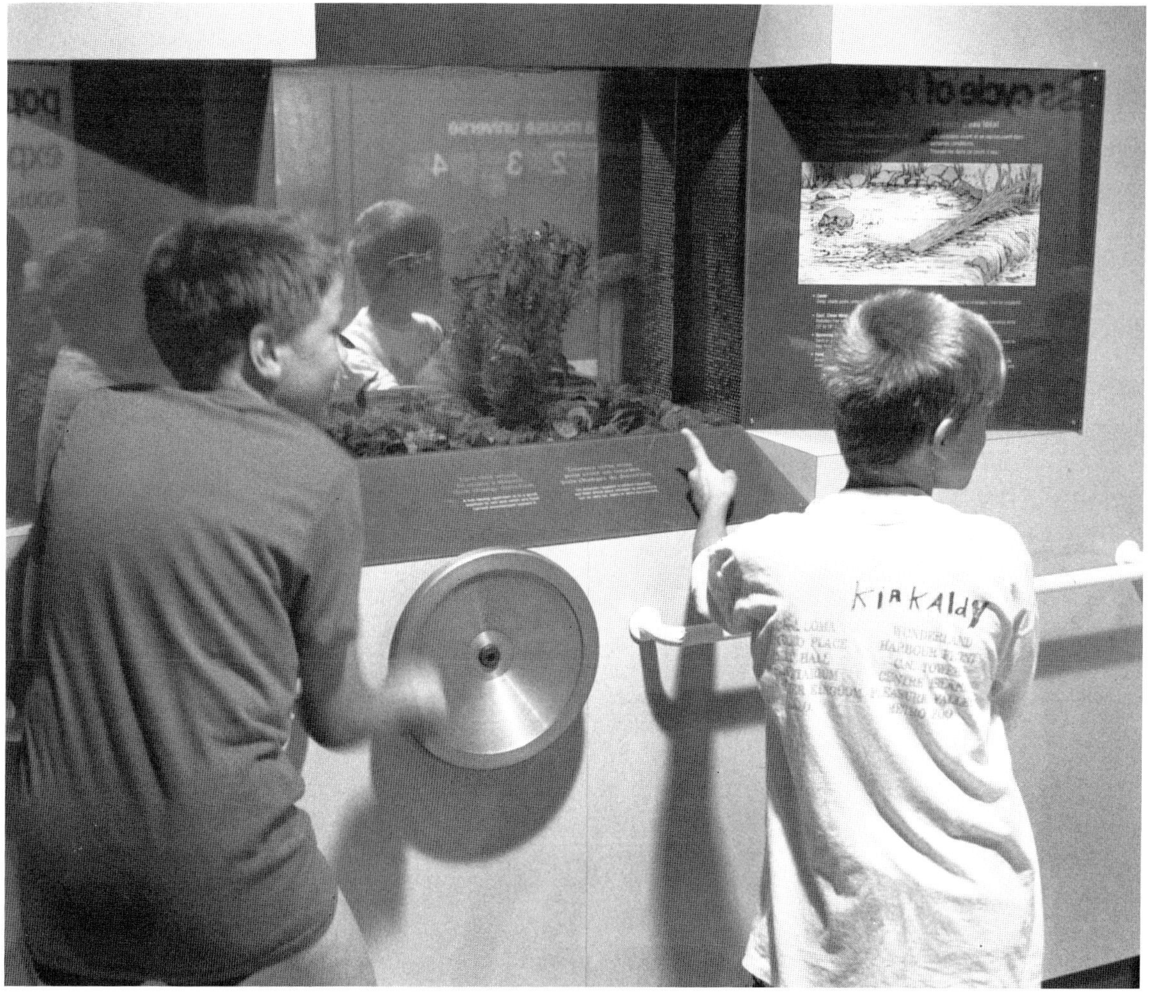

Large diameter handle This is a good approach when continuous rotary motion is required. The extended knob should spin freely on a shaft. Power and leverage are increased through the use of large diameter handles, like the one shown here. Additional power can be gained by reduction gearing between the handle shaft and the exhibit apparatus. The longer the knob, the more accessible the control becomes for people with motor control impairments.

> **SAFETY CONSIDERATIONS**
>
> 1. Handles and cranks used to transfer power to exhibit apparatus should always incorporate clutches so that energy cannot be transferred from the exhibit back to the user. A spinning knob with inertia behind it is a potential source of serious injury.
> 2. Handles which can be spun rapidly should not have spokes. Off-the-shelf industrial handles without spokes are described in product literature as "safety handles."

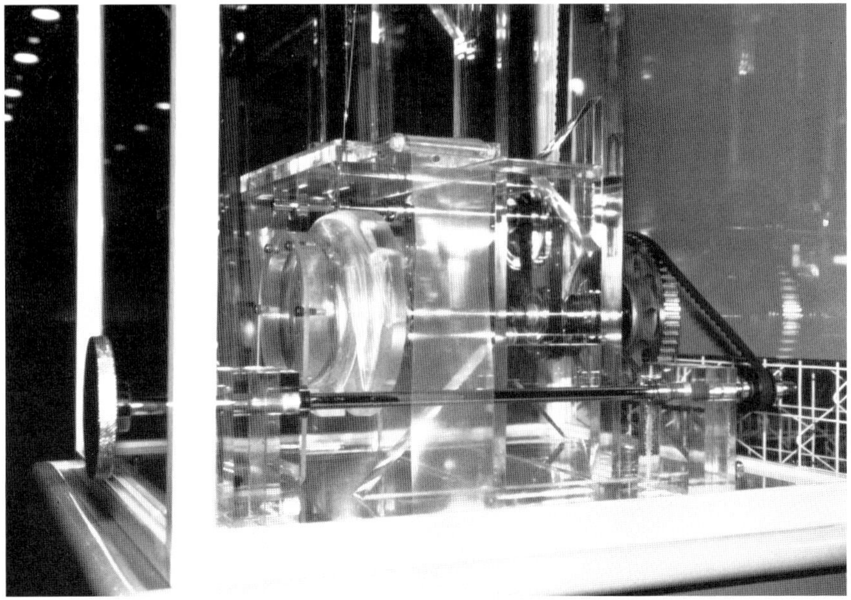

Round handle with knurled edge (no knob) This design is applicable to low-speed, non-continuous motion. It is also appropriate where precise movement is required. The diameter of the handle determines torque.

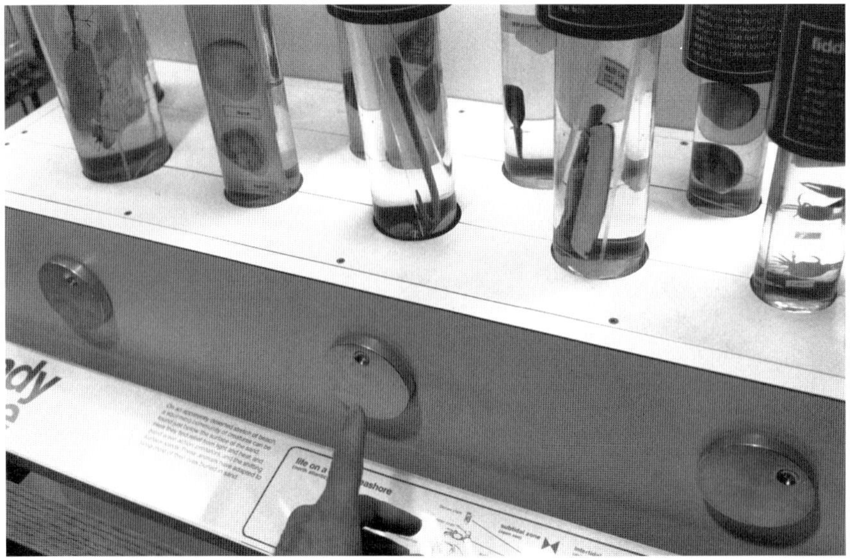

Round handle with finger hole This control minimizes the user's ability to go too fast and is typically used to limit the efforts of overzealous young users. However, this is a poor design for people with motor control impairments and should generally be avoided.

Small diameter knobs: fingertip controllers This approach is typically used when power needs to be restricted, or when little power from the user is needed. This, too, is a poor design for people with motor control impairments and should generally be avoided.

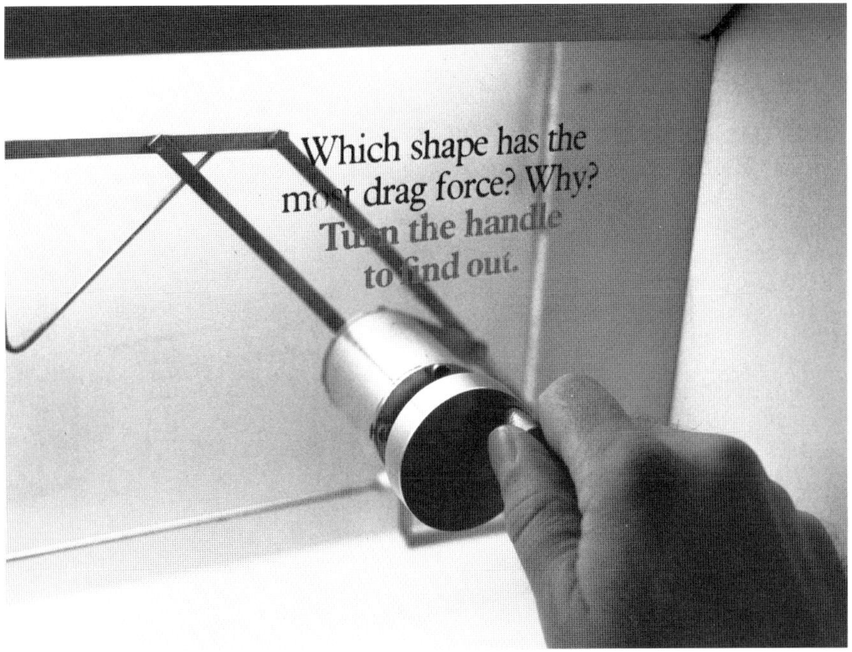

Levers Levers are useful when a great deal of power is needed. Here they provide the mechanical advantage needed to move a heavy weight.

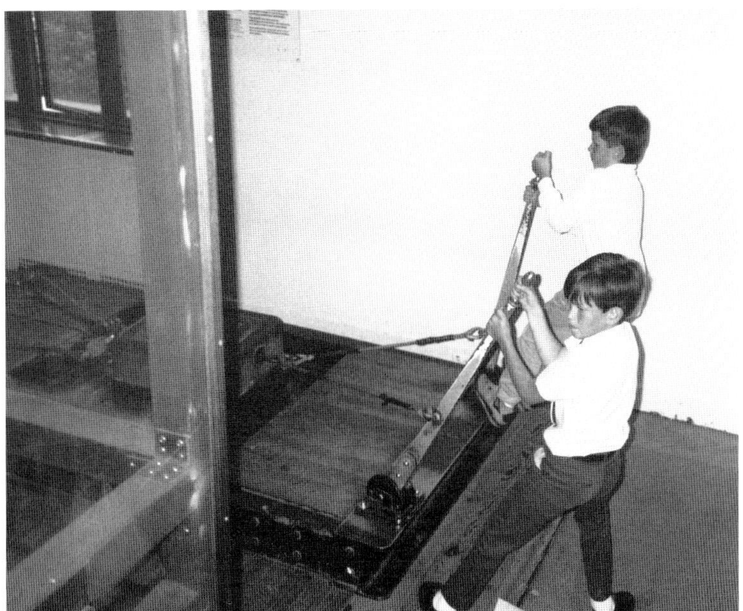

CONTROL SELECTION QUESTIONS

The following questions need to be addressed when designing mechanical controls:

1. What type of motion is required? (e.g., rotary, linear)
2. What is the range of motion required?
3. What is the desired range of operating speeds?
4. How much power is needed from the user?
5. How much precision is required?

Electrical controls

Electrical controls commonly used in hands-on exhibits include pushbuttons, joy sticks, trackballs, and an array of other devices for controlling the flow of electrons. The layout and labeling of electrical controls is one of the critical factors affecting a visitor's ability to use and understand the exhibit. Electrical control panels need to be evaluated during prototyping.

Pushbuttons Round pushbuttons are preferable to square ones, which have sharp corners and feel uncomfortable. The pushbuttons manufactured for the video arcade game industry are ideally suited to exhibits. These give a substantial "click" when depressed, providing feedback that contact has been made. They are also low cost, extremely durable, easy to install, and available in a variety of switching configurations and a range of colors.

Lighted pushbuttons are often used when an exhibit requires a sequence of operations. Lights provide the user with feedback about what actions have been taken and what to do next. When considering the need for lighted pushbuttons to help clarify a sequence of operations, it is probably good to ask whether the exhibit has become too complicated.

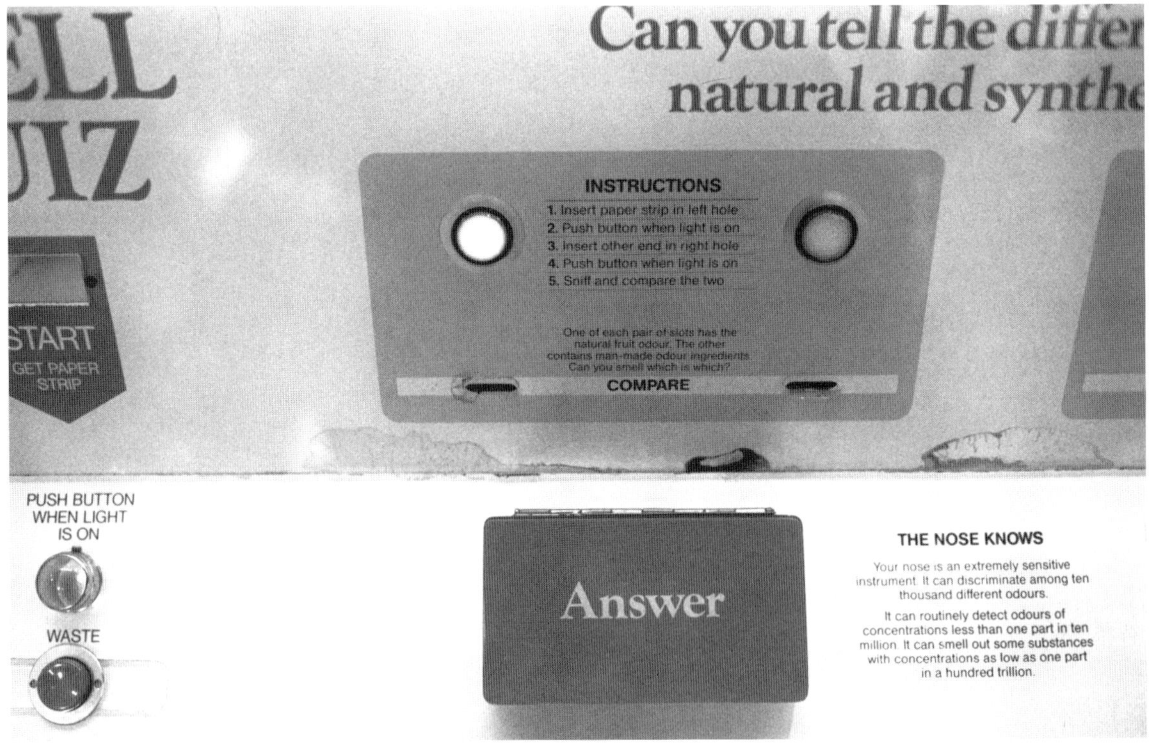

Trackballs A trackball is probably the most common device for controlling computer exhibits. Although they are simple to operate, trackballs are unfamiliar to visitors who have never used a computer. Here brief instructions have been provided to explain how to use a trackball.

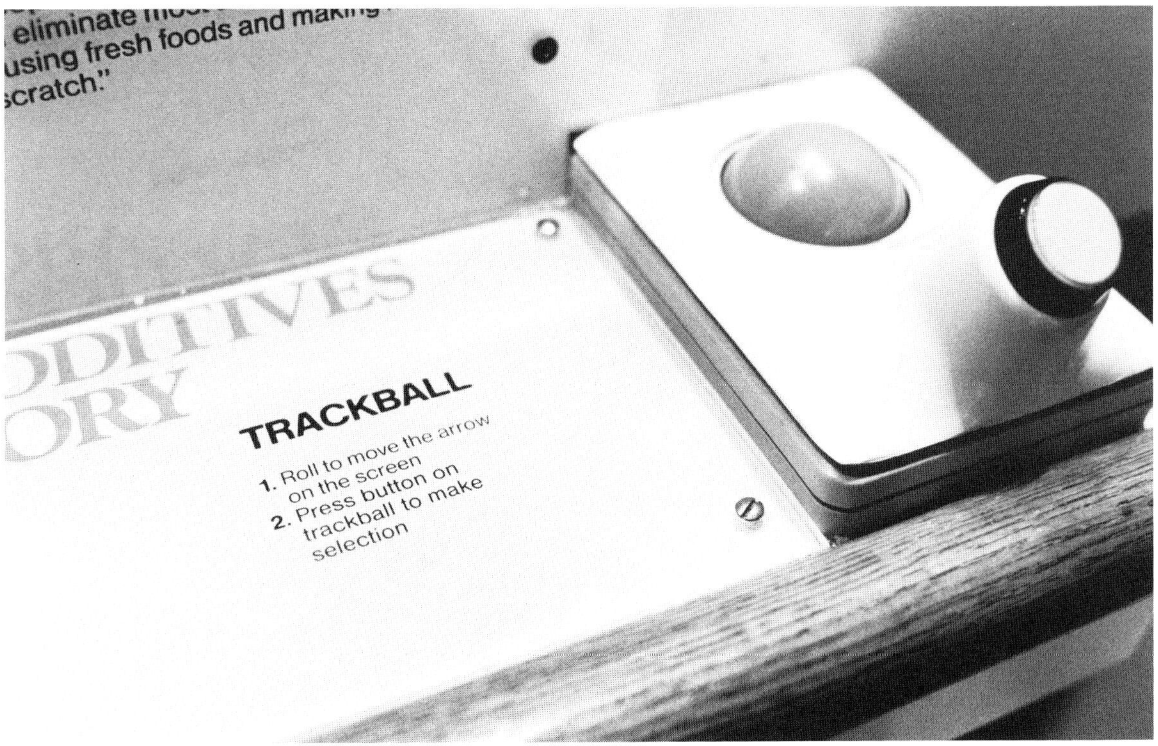

Joy sticks A joy stick is typically used to control multi-directional motion. In addition to directing cursors and other actions on video screens, joy sticks operate remote-control, multidirectional mechanical devices. Typical applications are robotic arms and video cameras that move in two or three axes.

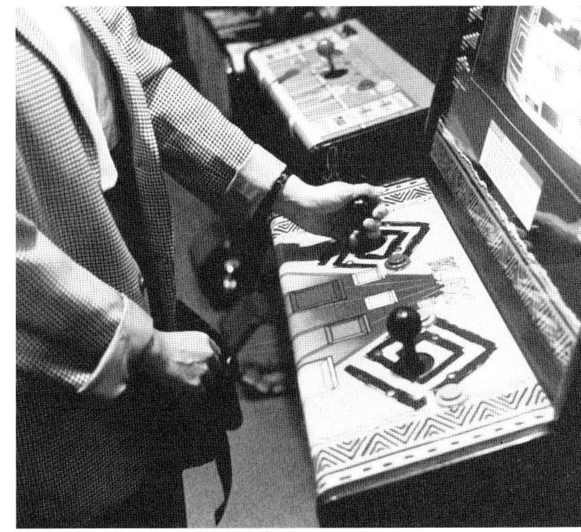

Two generations of controls
These control panels are for two identical exhibits. The earlier version (top) illustrates several small, easily correctable problems. First, making all the controls the same size and color and placing them in the same size circles obscures the fact that each has a different function. Second, the dark switches, placed against the dark background circles, are difficult to see. Last, the toggle switch (forward/reverse) and turning knob (disc speed) are too small to be used by people with motor impairments, and uncomfortable for other users as well.

The revised design (bottom) addresses each of these problems: the control for each function is boldly different from the others; there is a high level of contrast between controls and background; and both the pushbuttons and the turning knob are large and easy to use. In addition, the pushbuttons have internal lighting to indicate the state of the exhibit.

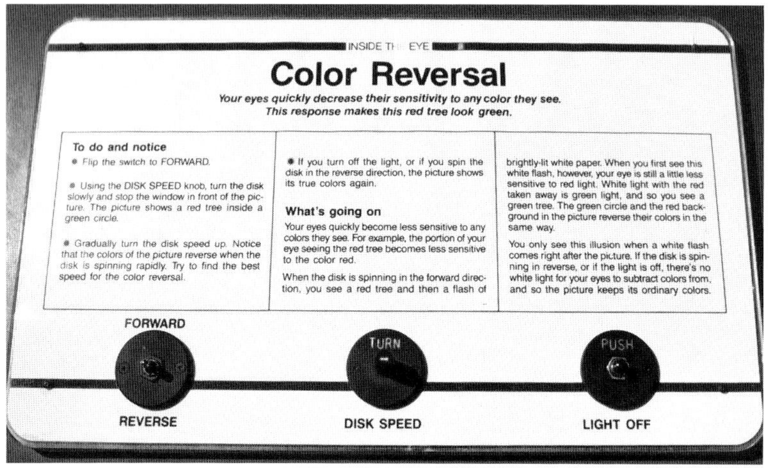

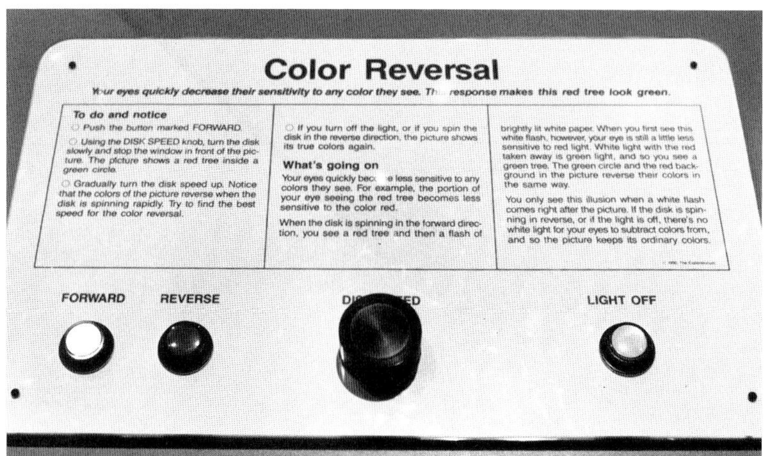

NATURAL MAPPING

Natural mapping takes advantage of physical analogies or cultural understandings. Natural mapping can be applied to the design of controls including handles, knobs, and pushbuttons.

Example 1: Physical analogy
An exhibit requires the user to control 25 lights arranged in a 5 x 5 grid. Employing a physical analogy, the control panel would be comprised of 25 pushbuttons arranged like the lights in a 5 x 5 grid .

Example 2: Cultural understandings
Pervasive cultural understandings in the Western World are derived from the convention of reading from left to right and top to bottom. Based on this convention, the following cultural understandings apply to the design of controls: left to right (clockwise) movement always means an increase; to cause forward movement, controls are moved up or to the right.

Eye Pieces and Viewers

4

Many hands-on exhibits require users to look through eye pieces or viewers – most often microscopes, telescopes, binoculars, and stereo viewers. These exhibits must provide for variation in visitor height and for comfortable and safe viewing.

How the design of an exhibit deals with varying visitor heights depends both on the phenomenon being demonstrated and on the specific interaction required. Based on our survey of eye viewer exhibits, there are four common solutions to varying user heights. These are illustrated and described in the following examples.

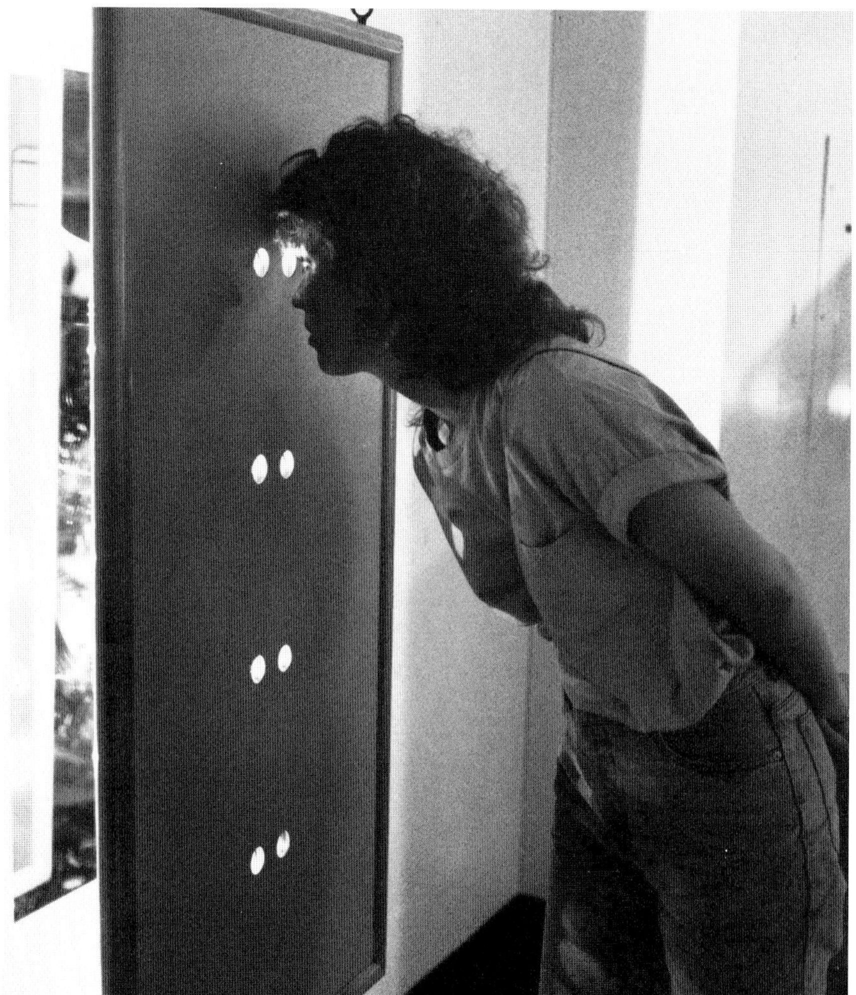

Multiple-height viewing ports This exhibit needs to be viewed through two eye holes, but does not require viewing from a single height. A simple, effective design solution is to provide eye holes at different heights for different size visitors. The four heights provided here are 32, 42, 50, and 60 inches from floor to eye hole center line.

Pull-out viewers This exhibit asks visitors to look at several sets of contrasting colors through a selection of viewers. The vantage point of the user is not particularly critical. A simple, effective design solution is to mount the viewers on retractable cables so that they can be brought to the eye of the user. This obviates the problem of user height variation.

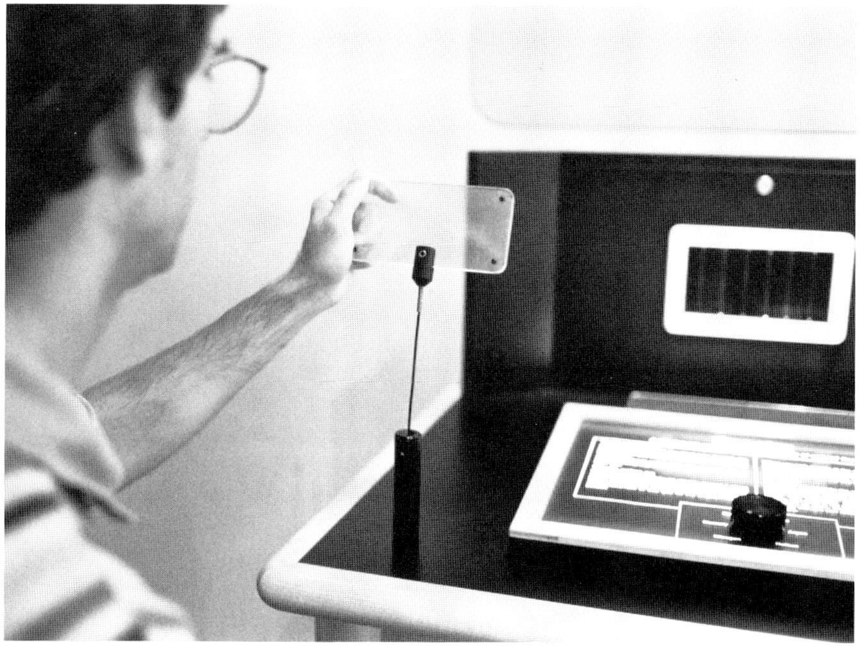

Components at different heights Here, identical exhibit components are provided at different heights. This is probably the most expensive solution to accommodating diverse user sizes. One advantage to this approach, however, is that several people can use the exhibit at once. This visitor is not using the component best suited to his height.

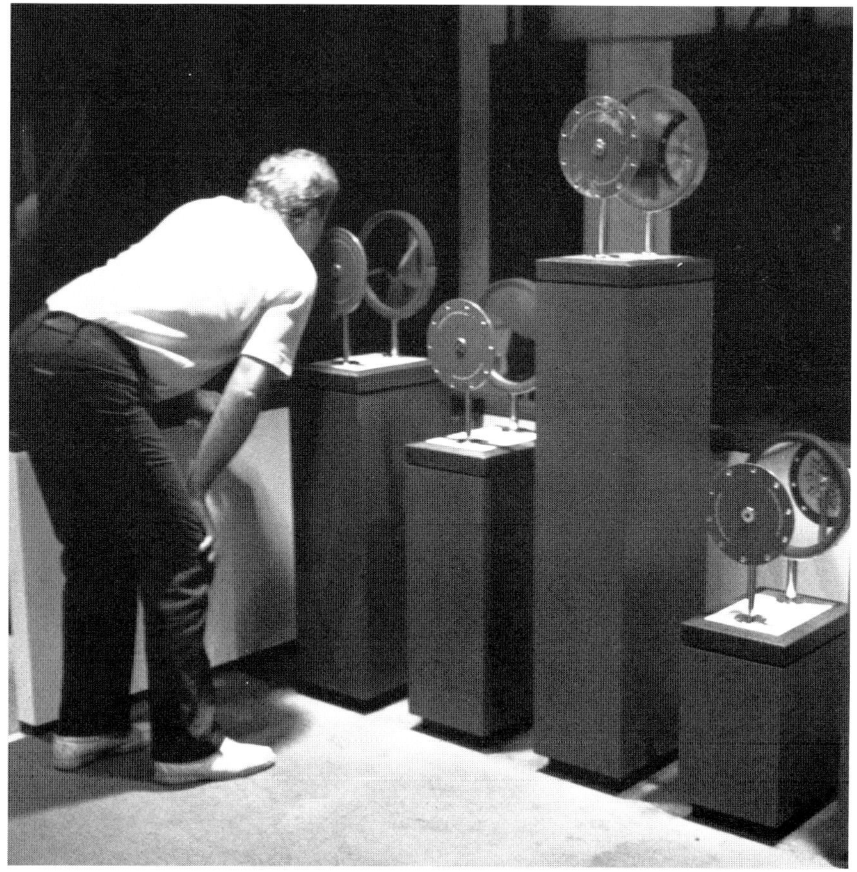

Seating used as a height equalizer
This example shows how seating can be used to deal with varying heights in a non-tabletop exhibit. The distance between the chair and the exhibit, and the angle of the user's back, are the two variables that change to accommodate different sizes.

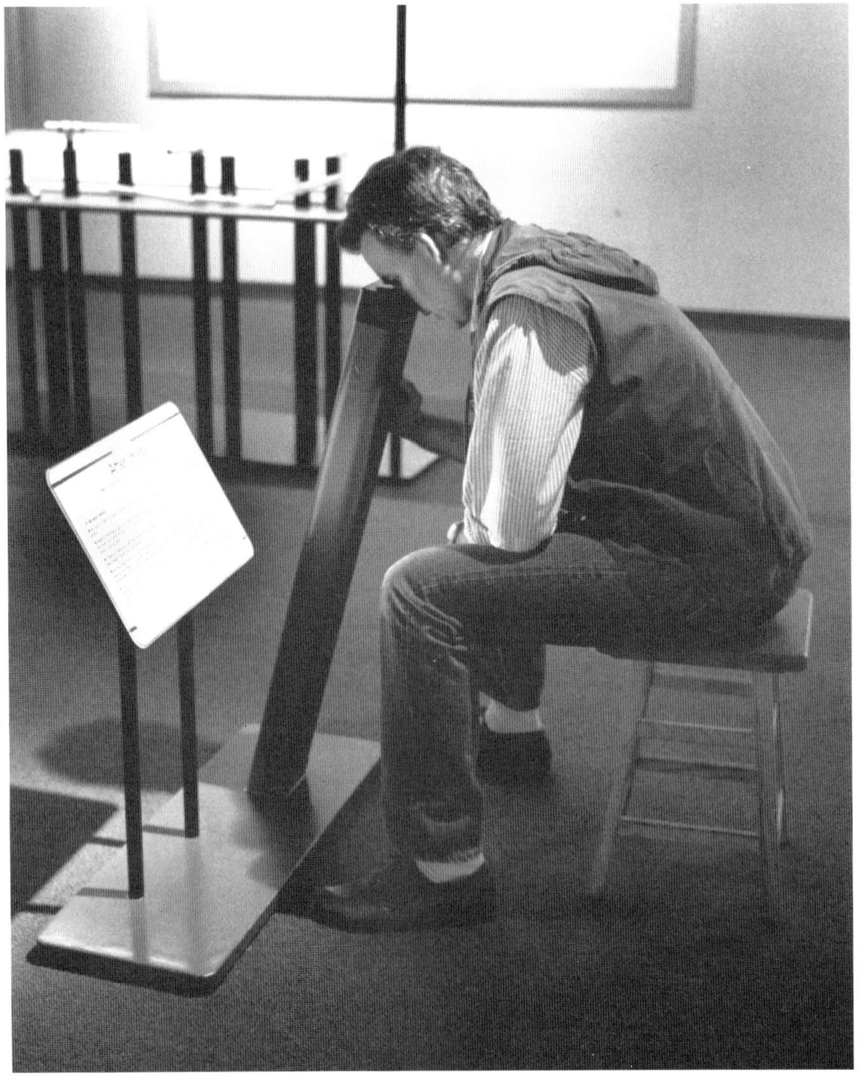

Factors contributing to the comfortable use of an eye viewer include the physical form of the eye viewer, the material from which it is constructed, and mounting details.

Physical form and materials

The primary requirement for the physical form of an eye viewer is that it be free of sharp edges. In fact, the softer and more "friendly" the viewer looks, the more comfortable the user will be when placing it against his or her face.

The material from which the viewer is constructed is also an important consideration. Metal can be machined to desirable forms, but feels cold and hard. Plastic is probably the best choice of materials for most applications. It is readily machined, thermo-formed or cast, and feels warmer and softer than metal. It is also durable and can be cleaned easily. The following three viewers are comfortable to use.

Clearly a viewer The mask-like configuration of this viewer invites visitors to put their faces up to it. All corners are generously curved and top edges are rounded over so that it looks and feels comfortable.

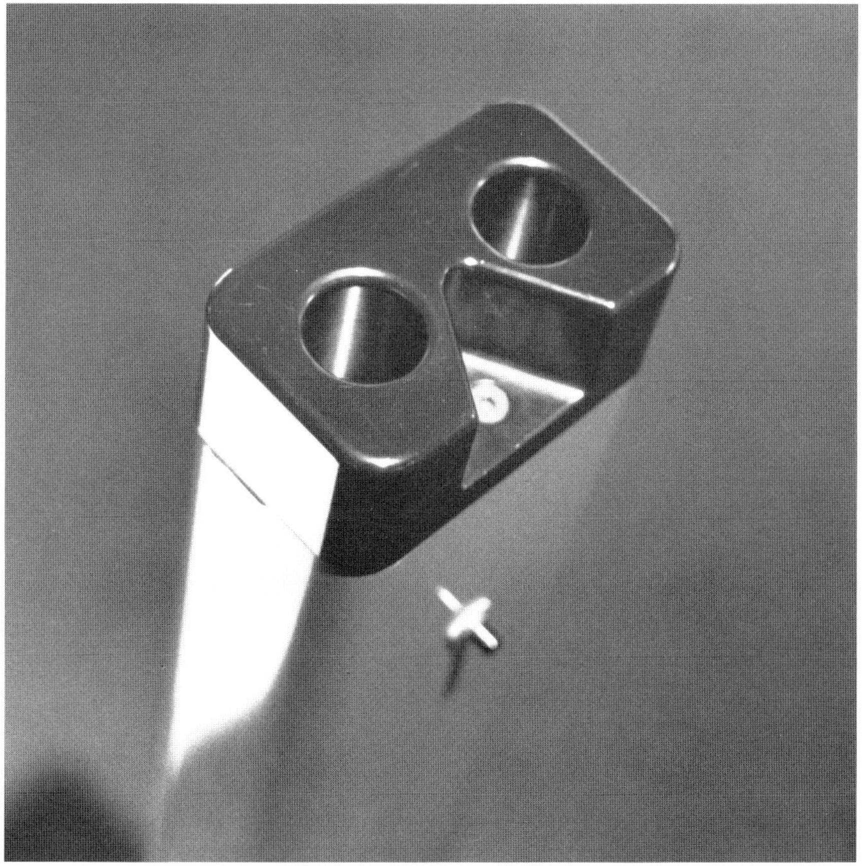

Hand-held peephole lenses These viewers are made of both cast and machined plastic parts. The grip is shaped to fit comfortably in the hand and has a friendly feeling.

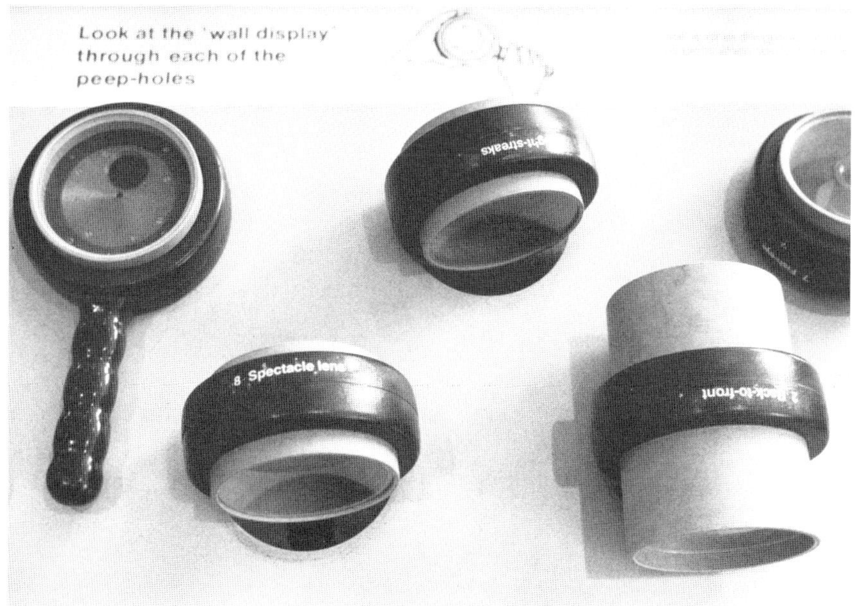

Hand-held viewer The corners of this viewer have been cut at a 45-degree angle to make it easy to hold. The plastic viewing lens has corners sufficiently rounded to make it safe to use around the eye. The metal handle lacks the "friendly" finger grips of the peephole lenses and feels less inviting.

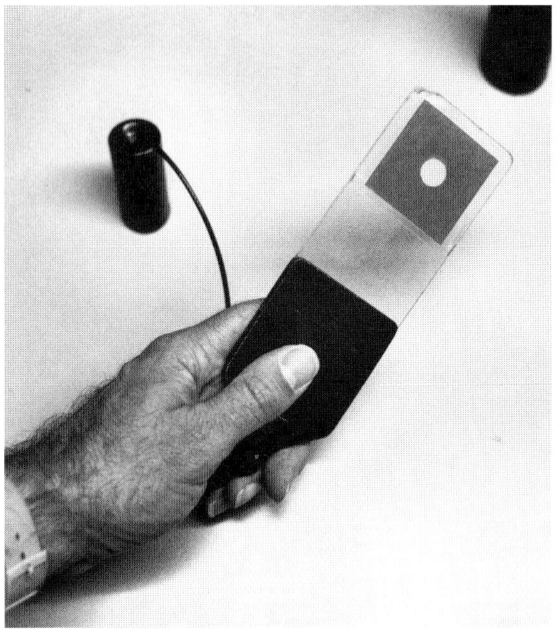

MICROSCOPE VIEWER ALTERNATIVES

Microscope viewers are very useful interpretive tools. Unfortunately, laboratory-style microscopes and stereoscopes are quite difficult for inexperienced operators to use and frequently become a source of visitor frustration. A number of alternatives are available for showing magnified images without looking into an eye piece. The exhibit "Hidden Kingdoms: The World of Microbes" at the New York Hall of Science uses four different kinds of microscopes, none requiring eye-piece viewing.

1. *Buxton Micrarium projection microscope*
   The image of live specimens is rear-projected onto a plastic screen. Magnification is up to 140x. Users can select from four specimens, use a joy stick to move about, zoom in and out, and focus.
2. *Easy View microscope*
   The image is viewed in a four inch diameter display. Magnification is from 20x to 100x. Users can move specimens under the lens. This instrument was developed specifically for museum and school use and is extremely user friendly.
3. *Video microscope*
   A video camera is attached to a laboratory microscope and the image of living specimens is displayed on a large video monitor for group viewing. Magnification is from 40x to 1000x. This exhibit is operated by demonstrators only.
4. *Scanning electron microscope*
   This instrument is not for viewing live specimens. The image is displayed on a large video monitor for group viewing. Magnification is from 35x to 300,000x. This instrument is operated by demonstrators only.

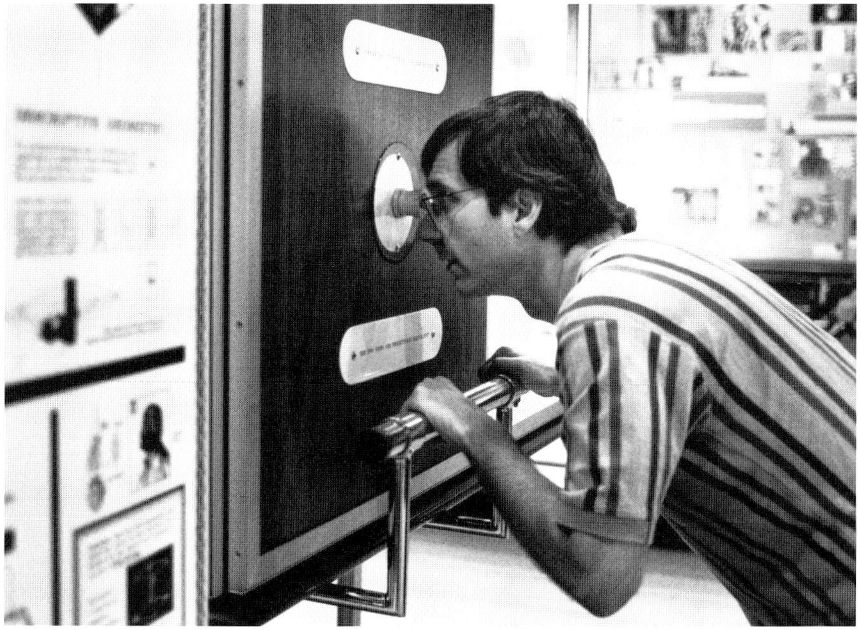

Mounting details Even the way a viewer is mounted can add to user comfort. An interesting example is the addition of a hand grip next to the eye piece. Almost without exception, visitors hold on to this grip when looking into the viewer. The grip serves two purposes. First, users can brace themselves in case passers-by bump into them. Second, they can use it to anchor themselves and to steady the head and eye.

5

Interactive Video and Computers

Interactive video and computers, with their responsiveness to individual users, sophisticated motion graphics, and integrated audio, are powerful tools for interpreting complex concepts in exhibits. Of all of the types of hands-on exhibits, however, interactive video and computers present the greatest number of human-operator design problems. These range from the design of the user station to the design of screen navigation.

Presentation formats for interactive video and computers

Interactive video and computer-based exhibits are most often presented in one of three formats, or a combination of formats. These include:

Selective access video programs

A key feature of videodisc technology is that the laser can access any location on the disc almost instantly. Responding to this capability are exhibits that let users selectively access video program material by pushing a button or touching a monitor. This approach can be used to break long productions into short chapters which visitors browse through according to personal interest and schedule. This is the simplest form of interactive video, but has extensive applications and is used widely. On-screen captioning is easily provided on videodisc and should be included with all video presentations. Captioning benefits visitors who have hearing impairments and those not familiar with (or fluent in) the language. In noisy environments, captioning is useful for all viewers.

Hybrid interactive video and interactive computer exhibits

Developments in computer graphics software, control software, and mass storage have led to a generation of complex, highly interactive, computer-controlled video exhibits. Combining features of film, animation, and computer graphics, these exhibits provide designers with sophisticated tools for developing games and simulations that are accessible to users of varying skill and knowledge levels. Creating a complex program is expensive and requires substantial time for planning, development, and production. The biggest problem posed by complex interactive video is the design of a usable and understandable visitor interface.

Interactive computer graphics

Animated computer graphics have also emerged as a powerful tool for interpreting complex concepts through interactive games and simulations. Although computer illustration and animation software is steadily improving, ambitious projects require a great deal of illustration and programming time, even by the most skillful practitioners. Like interactive video, the most challenging design problem is creating a usable, understandable visitor interface.

The design of interactive video and computer exhibits varies greatly depending on the number of users for which it is intended. The most common form of interactive video and computer exhibit is the single-user station, or a cluster of single-user stations. However, there is a growing interest in group interaction that is being fueled by decreasing costs in control system hardware and software. Following are examples of human-operator issues for presentations in a variety of size formats.

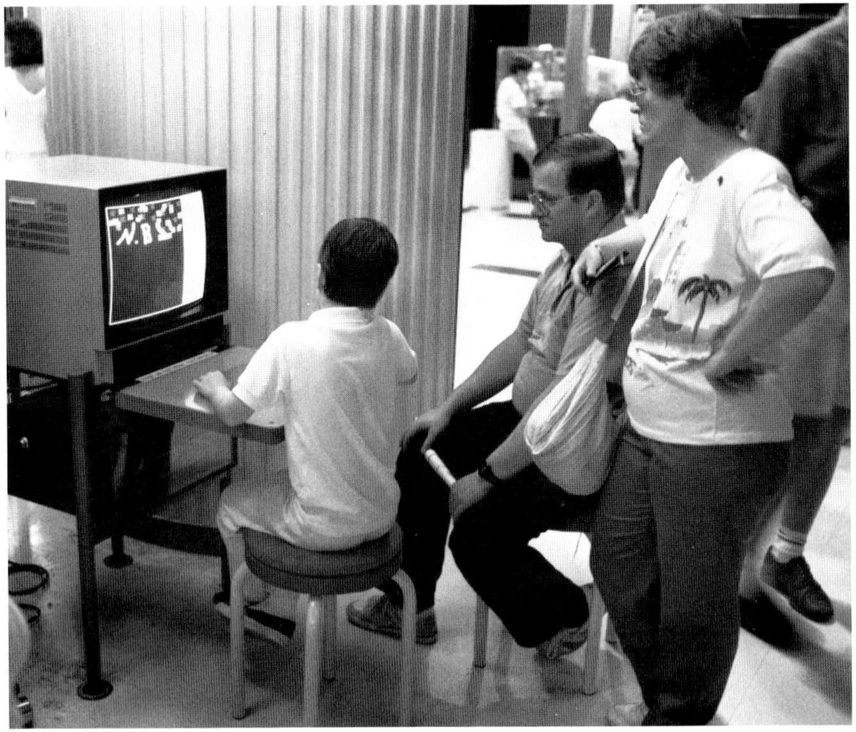

Single-user station
Although the single-user station is operated by one person, other people usually look on. This reflects the fact that museum visitors usually come with friends, family, or a group. A parent and child often use a station together. The stations shown here have stools to accommodate two people.

Repeater screens Another way of accommodating onlookers is to provide a large auxiliary viewing monitor positioned high enough to be seen from a distance. These are sometimes referred to as "repeater" screens since they are not interactive. If auxiliary audio is also provided, sound spill may pose a problem.

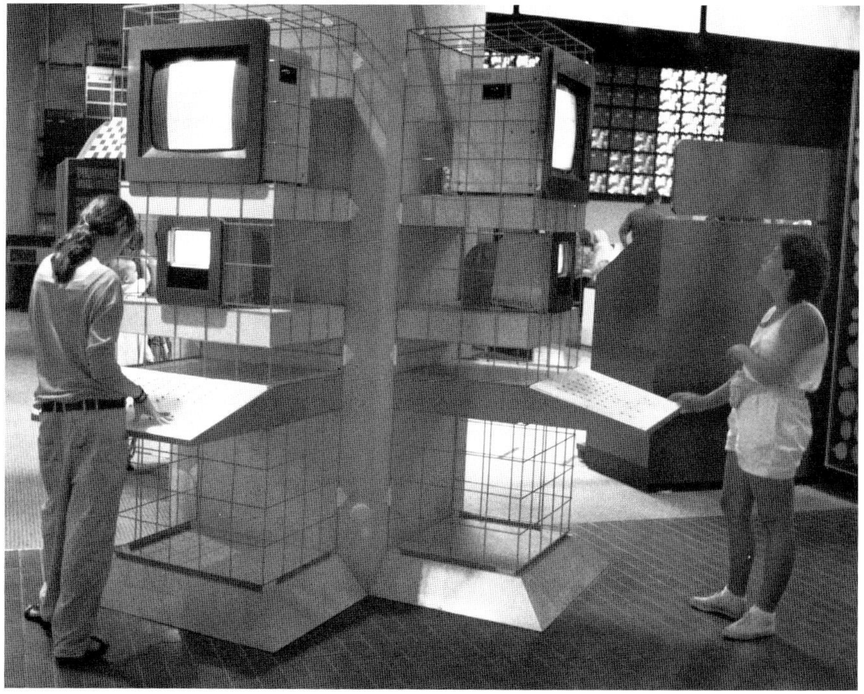

Multiple stations Visitors may spend as much as 10 minutes at a highly interactive video or computer exhibit. To serve more visitors, multiple stations of the same computer activities have been provided here. These particular stations feature programs for young users and have lower seating than usual (15 inches).

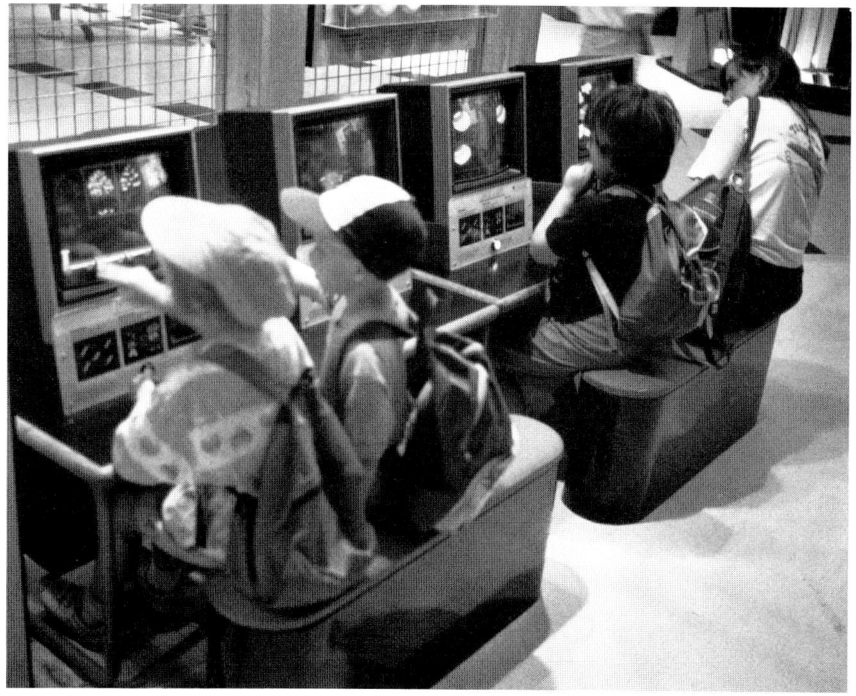

Small group interaction Visitors play the role of Olympic judges in this exhibit with six interactive voting stations. The semicircular layout of stations helps to create a feeling of being part of a group. This is also a good configuration for accommodating a large number of onlookers. The stations are accessible to wheelchair users.

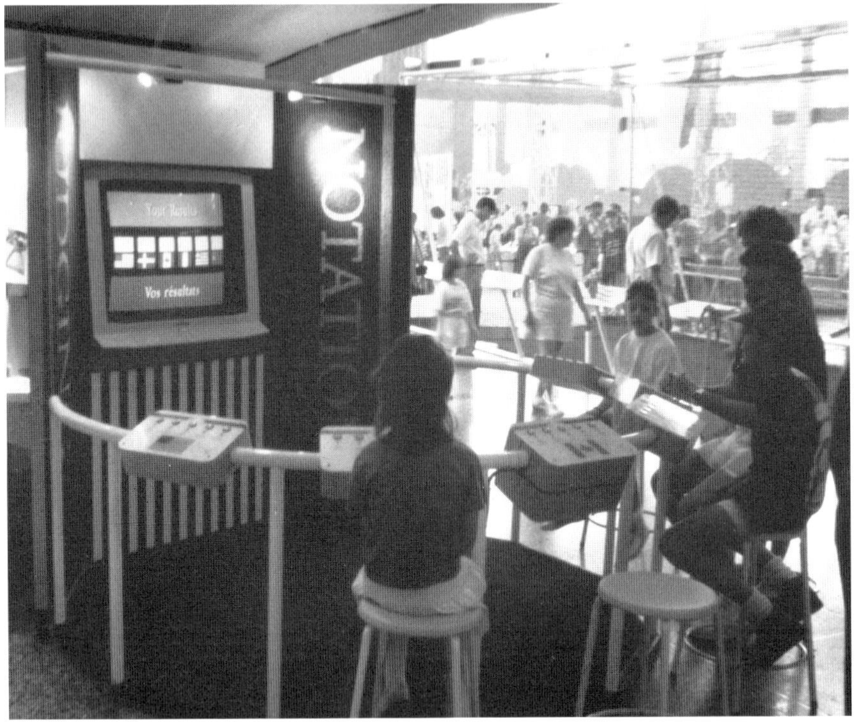

Large group interaction This semi-enclosed theater seats 72 people. The lines of sight are tricky in this tight space, but have been well resolved. Seating is very comfortable. Bright daylight often shines over the top of the theater walls, however, causing a serious distraction. Visitors participate in a multiple-choice quiz by pushing buttons on a pad in front of their seat. Results are tabulated and rear-projected onto large screens.

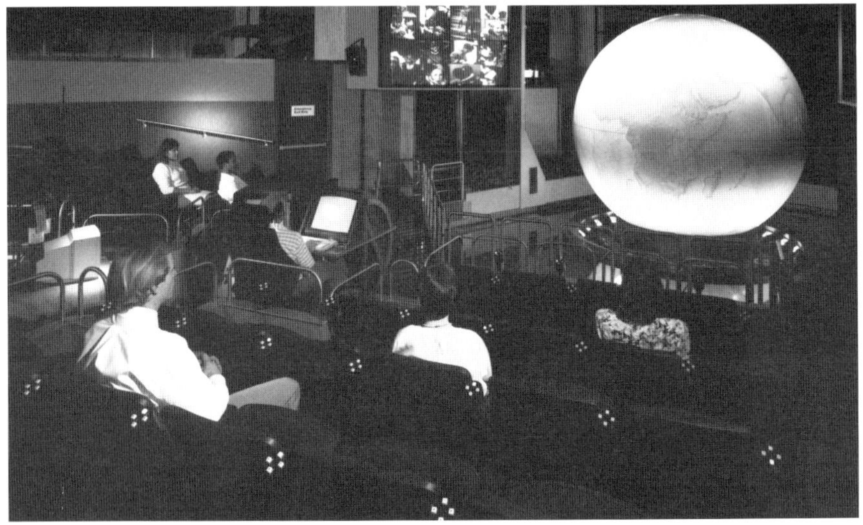

One user with a large audience This 108-monitor interactive video wall has a video camera and keyboard that visitors can use to zoom, take pictures, and display pictures in various sizes on the wall. The user area provides good seating, lighting, usable controls, and a comfortable, expanded rest area for onlookers and those waiting to use the exhibit.

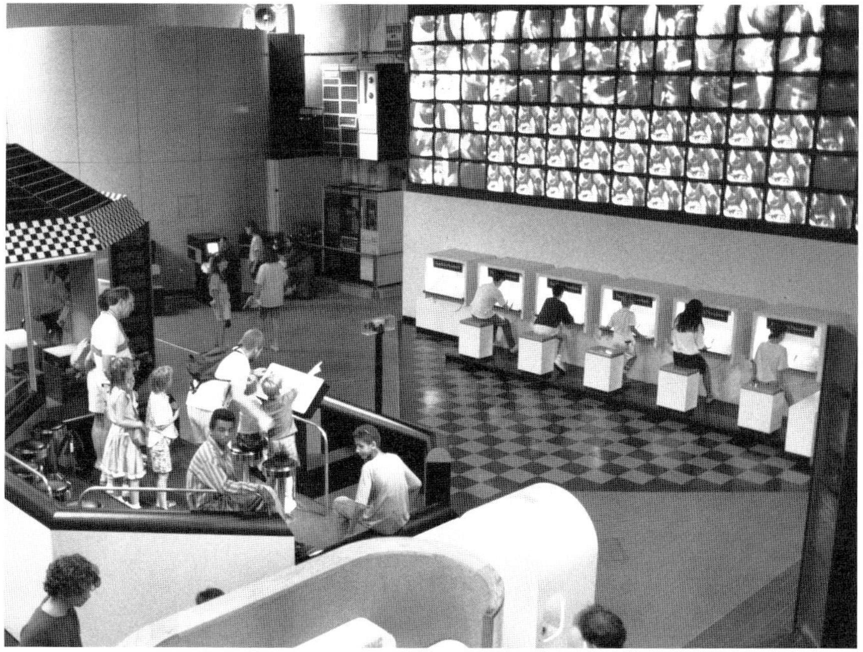

USER FRIENDLY CHECKLIST

A user friendly interactive video or computer exhibit is one that visitors can use and understand easily, and can interact with comfortably. The following checklist suggests the range of issues involved in designing an exhibit that meets these criteria.

1. Does the visitor understand what to do?
2. Is the goal of the activity clear?
3. Is the input device (a touch-screen, for example) appropriate for the required task?
4. Does the user understand how to use the input device? Is it comfortable to use?
5. Are the choices for actions understandable?
6. When an action is taken, is there immediate feedback, informing the user of the results of his or her action?
7. Is there a clear escape route, allowing the user to return easily to any previous point in the program at any time?
8. Are response times quick, or does the user spend a lot of time waiting for the system to perform operations?
9. Does the user have to give a disproportionate amount of time and attention to the task of system navigation?
10. Does the physical design of the work station allow comfortable use by visitors of all sizes and by wheelchair users?
11. When instructions are conveyed through audio, are there captioned instructions to provide access to deaf or hearing-impaired visitors? Similarly, when instructions are conveyed through screen text, is there an audio track for users with sight impairments?

Visitors interact with interactive video and computer exhibits using an input device. With the advent of menu-driven software, keyboards have largely disappeared and have been replaced with either pushbuttons, touch-screens, or trackballs. Joy sticks are also used with exhibits requiring greater spatial control. Young people are more familiar and more comfortable with trackballs, touch-screens, and joy sticks than are adults.

Pushbuttons

Pushbuttons are probably the best choice for simple exhibits giving visitors selective access to video material. Pushbuttons are inexpensive, reliable, and easily replaceable. When pushbuttons are large, easily depressed, and well spaced on a panel, people with motor control impairments can operate them easily.

Touch-screens

Touch-screens are applied to the front of a computer or television monitor and are invisible to users. There are several touch-screen technologies in use, but all work on the principle of dividing the screen into a grid matrix. Graphic "pushbuttons" are displayed on the monitor in alignment with the grid. When the user touches a graphic pushbutton, a corresponding circuit on the touch-screen is activated. The use of the finger as the primary controller means that the resolution of the grid matrix is quite coarse, especially in comparison with a trackball cursor. A plus, however, is that by touch-activating the graphic pushbuttons, the user is provided with a direct feeling of control. It is essential that pushbuttons provide visual (blinking, color change) and/or auditory (clicking, beeping) feedback to let users know contact has been made. Touch-screens are relatively expensive to replace when they wear out.

Because touch-screens carry a lot of information, design of the screen is a critical part of an exhibit's appeal, usability, and comprehensibility. For this reason, screen design is quickly becoming a specialized area of interactive video design.

Trackballs

Trackballs are widely used in exhibits. Rolling the ball controls the movement of a cursor on the computer or television monitor. Some exhibits are designed so that moving the cursor on to an "action icon" initiates an action. Others have a pushbutton next to the trackball. After moving the cursor on to the desired icon, the button is pushed to initiate an action.

As with touch-screens, trackball-based screen design is a critical aspect of the exhibit and is often the weak link in an otherwise well-conceived project.

Although manipulating a trackball requires greater hand-eye coordination than the touch-screen, this does not seem to present a problem for most users. Many museums have chosen trackballs over touch-screens because of their lower cost and greater durability.

Joy sticks

A joy stick is like a trackball with a lever attached for the user to hold on to. This type of control is popular for home computer video games and is therefore familiar to a sizable segment of the younger audience. The grip often has a built-in pushbutton which makes one-hand operation very easy. Because of the leverage provided by the grip, joy sticks need to be designed to be exceptionally rugged.

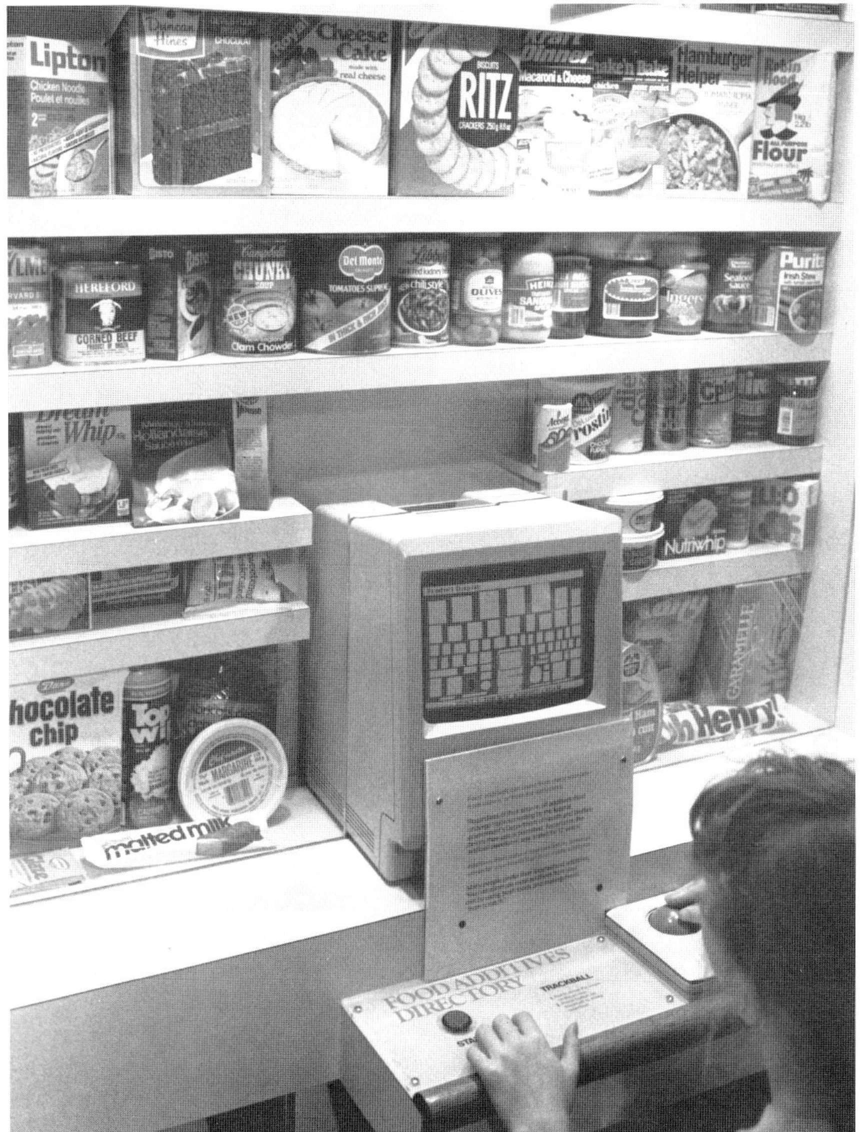

Potential and pitfalls
With the cost of graphics-based computers dropping and the tools for creating interactive programs being simplified, more small computer exhibit stations have appeared. Although these offer great potential, there are pitfalls as well. One is that these systems run very slowly unless expertly programmed and used on high performance equipment. Also, text on small screens can be very difficult to read.

As a rule, interactive video and computer exhibits should be designed to be operated while seated if long interaction is desired. Seating adds greatly to user comfort and there is evidence that visitors spend more time interacting with video exhibits when seating is provided.[3] A design with seating also makes these exhibits accessible to wheelchair users, if the seating is easily movable. The examples provided here reveal a range of design problems associated with single-user interactive video and computer stations.

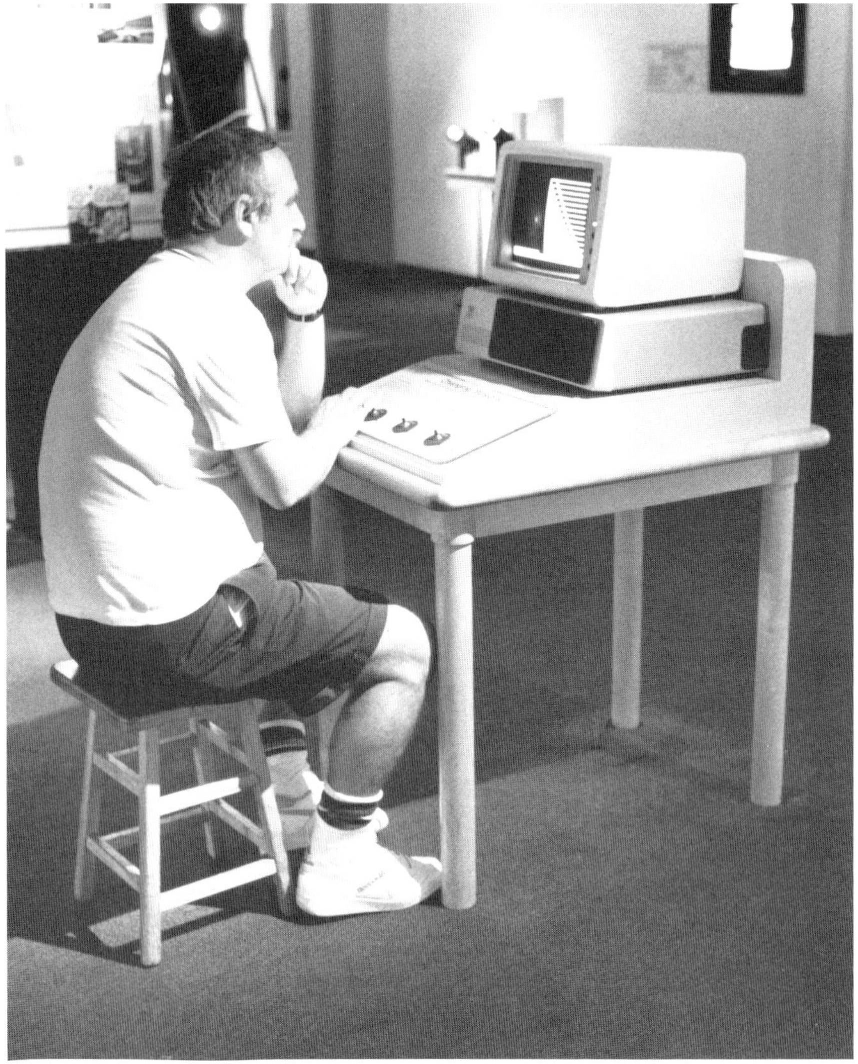

Computer-on-a-table
This is as simple and nearly as effective as it gets. The computer sits on a table high enough to accommodate a standard wheelchair (30-inch clearance under tabletop). The control panel is mounted at a good angle for viewing. All edges of the table are comfortably rounded wood.

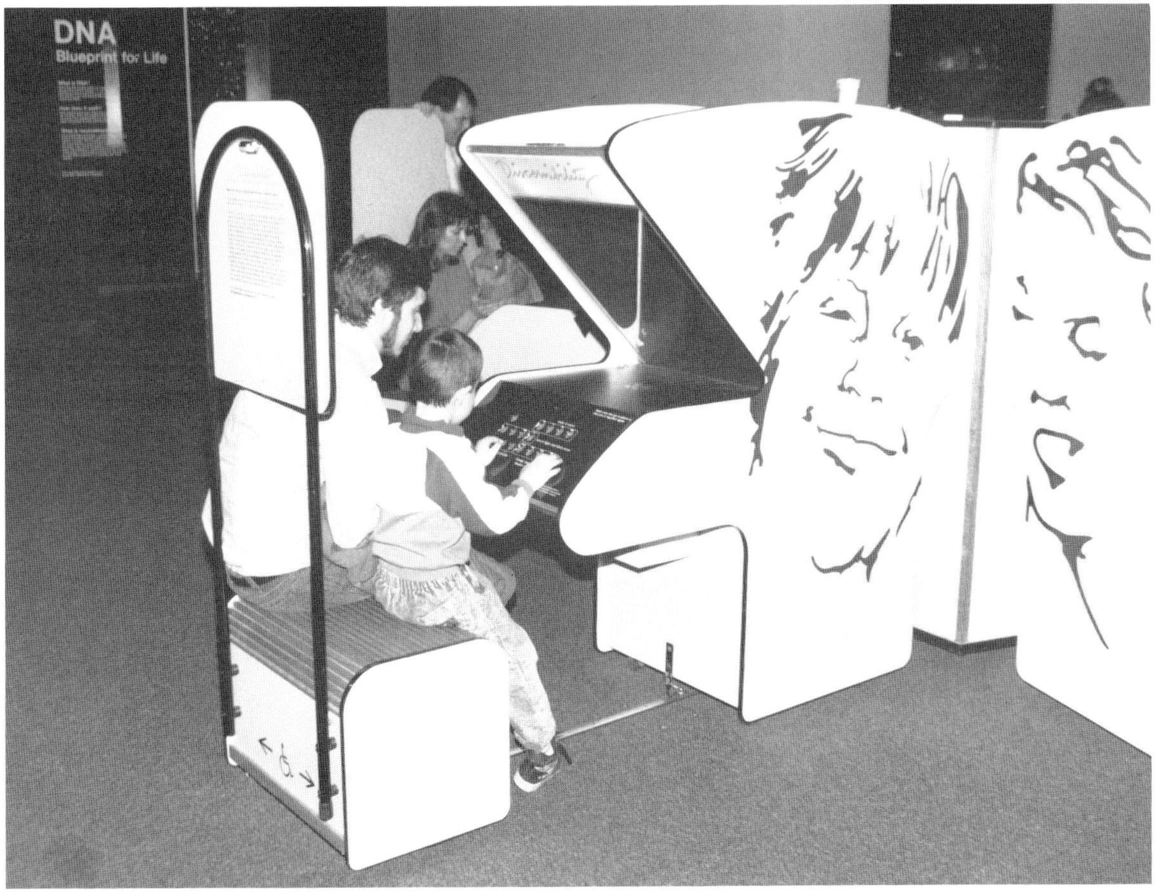

Complex work station This exhibit shows careful attention to many typical computer station design problems. Screen glare is controlled with a hood over the screen and a black, non-reflective surface in front of it. Glare is distracting to all viewers and is particularly uncomfortable for viewers with visual impairments. Glare control is especially needed in traveling exhibits, like this one, which encounter unpredictable lighting situations.

The control panel is set at an angle for comfortable viewing and operation. All corners of the housing are curved and softened with vinyl molding. The backboard attached to the seating provides a sense of privacy, as well as functioning as a text panel. The seating is wide enough to accommodate two people – a bonus for a parent and child. Perhaps the most interesting feature of this design is the roll-away seating that was designed specifically to accommodate people using wheelchairs.

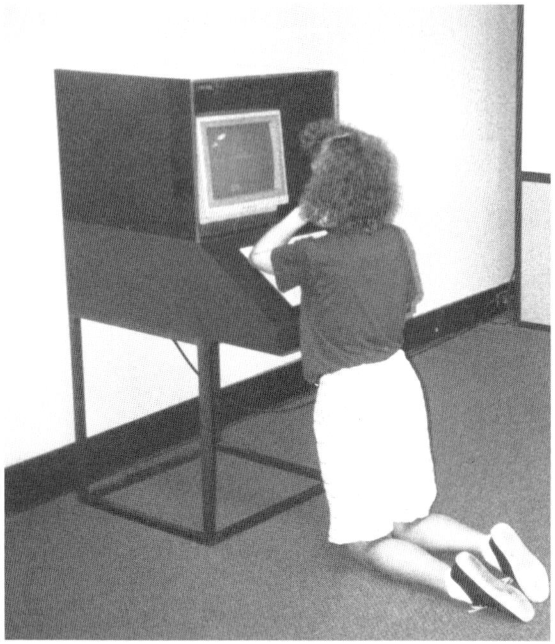

Missing The problem with movable chairs is that they sometimes move too far away.

These diagrams provide dimensional guidelines for a single-user interactive video or computer exhibit. The criteria for these guidelines are:

- The exhibit should accommodate a wide range of users including a five year old, an average size adult, and a wheelchair user.
- Wheelchair access should be from the front rather than from the side.
- Control panels and text panels should be angled to make them easier to read from a seated position.

One size doesn't fit all The difficulty of locating a touch-screen monitor at a single height is evident here. Placed at a height comfortable for a standing adult, this monitor is difficult for small users to reach. It is also inaccessible to wheelchair users due to its height and because it does not allow frontal access.

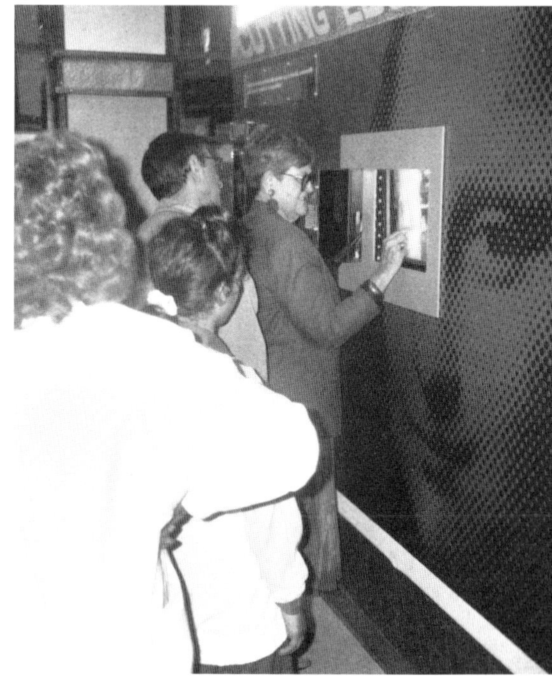

Using a touch-screen TV monitor while standing (top)This diagram shows a 13-inch TV monitor placed to accomodate a full range of users. The centerline height of 47 inches and 12-degree upward tilt were determined by averaging reach and sight line data of all users. For increased comfort, the TV monitor should be mounted so that visitors can adjust the angle to match their height.

Using a computer exhibit while seated (bottom) This diagram shows a common configuration for a computer or interactive video presentation. The 30-inch table height allows a standard wheelchair arm to clear the underside of the table. The 18-inch stool is a standard in most museums. The monitor height zone for comfortable viewing by all users is between 39 inches and 48 inches. Although the feet of five year olds dangle in the air, children can reach controls and have a good view of the screen.

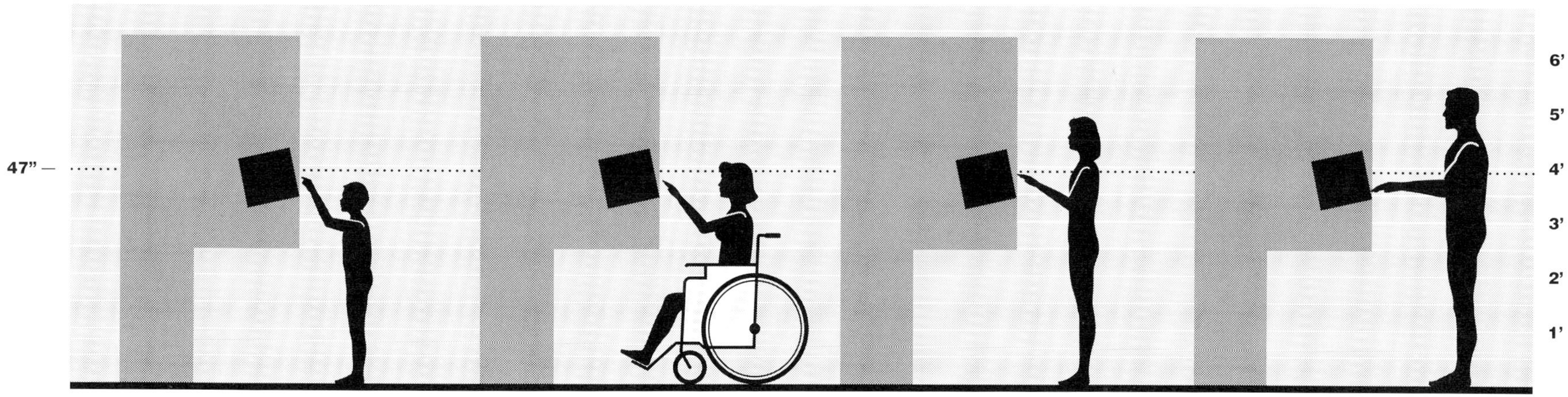

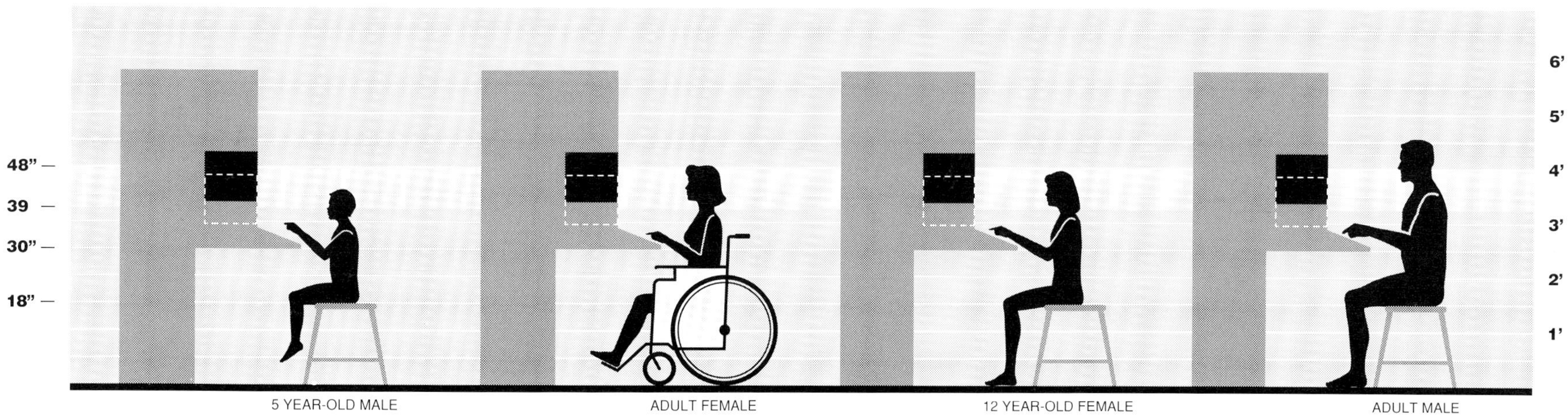

The need for prototyping and evaluation

Interactive video and computer exhibits have many allures: the ability to visualize abstract concepts; the ability to address topics in depth; the ability to simulate real and imaginary worlds; adaptability to the skill and knowledge level of the user. The flip side is that the user interface software and screen-design issues are extremely complex. This no doubt accounts for the incredibly high percentage of interactive video and computer exhibits that are unclear and frustrating and at best tedious to use.

In order to improve chances for a successful user interface, it is essential that the design process include rigorous prototyping and evaluation. The evaluation should use a public audience and should provide an on-screen experience as close to the final presentation as time and budget allow. Evaluation of an on-screen prototype will allow assessment of the navigation issues that are fundamental to the visitor's ability to use the program easily. When funding is not available for this level of prototyping and evaluation, mock-ups using static photos and a simulation of the program's decision-making process can be used. Testing and evaluation are doubly important because of the high cost of videodisc production. After all of the material for the program has been produced, edited, mastered, and the disc made, the cost of making changes is substantial.

6

Audio Exhibits

Audio is frequently a component of interactive video and computer exhibits. Sound pollution caused by the conflicting audio tracks from adjacent presentations is probably the most common violation of human factors design in museums today. Noise pollution makes hearing a presentation of choice difficult, and makes the task of concentrating on non-audiovisual exhibits in the same area extremely tiring.

Sound is typically presented in four ways: through a personal listening device; in acoustically isolated exhibit enclosures; in semi-acoustically isolated exhibit enclosures; and in open exhibit space without acoustical enclosure of any type. The pros and cons of each are examined in the following examples.

Headphones

Headphones provide users with the highest quality, private audio experience. In some exhibits, high quality stereo sound reproduction is extremely important to the interpretive experience. When volume controls are provided, users can adjust the sound level to their own hearing preference. For user comfort, the capability to adjust size is needed. Headphones present a public health concern because they rest on the user's hair. Another drawback is that headphones do not work well for visitors who use behind-the-ear hearing aids.

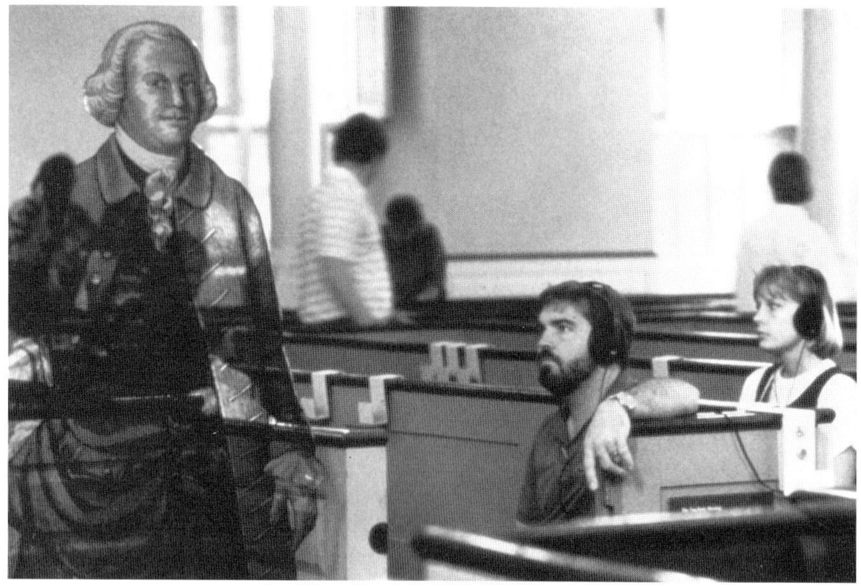

Headphones in a quiet setting Unfortunately, there are no headphones currently on the market that are built ruggedly enough to hold up in the science center environment. However, off-the-shelf, consumer headphones have been used successfully in exhibits where visitor behavior is more subdued or where there are seldom large crowds.

Monophonic ear speaker

The monophonic ear speaker, held up to the listener's ear, is probably the most common form of private sound delivery used in museums. Ear speaker devices include telephone handles adapted for this purpose and specialty products designed to meet the rigors of the museum environment. Telephone handles, like headphones, do not work well for visitors who use behind-the-ear hearing aids. "Hearphones" – the trade name of one of the most widely used products designed specifically for museums – seem to work for all visitors, including those using hearing aids.

The major drawback to the ear speaker is that ambient noise comes in the other ear, conflicting with the audio presentation. Another drawback is the very poor sound quality afforded by these devices.

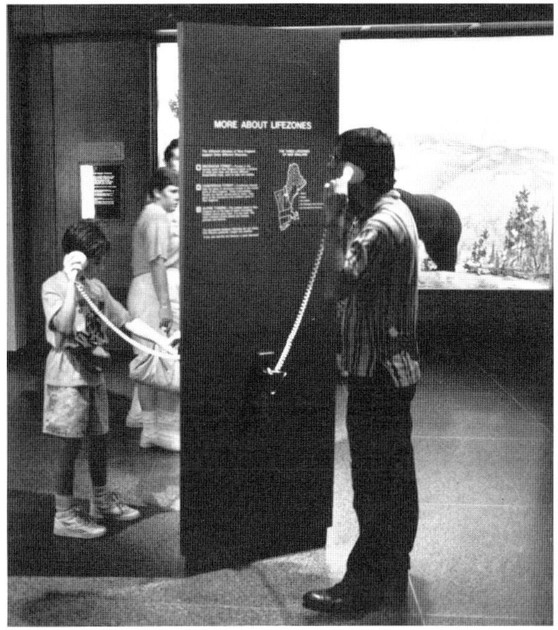

Noise control Regardless of its shortcomings, the ear speaker is one of the few ways of dealing with the problem of noise pollution in confined exhibit spaces. It is probably more desirable to have the drawbacks of the ear speaker than to have conflicting sounds permeating an exhibit area. This approach also makes it possible to provide audio labels for individuals with sight impairments throughout an exhibit.

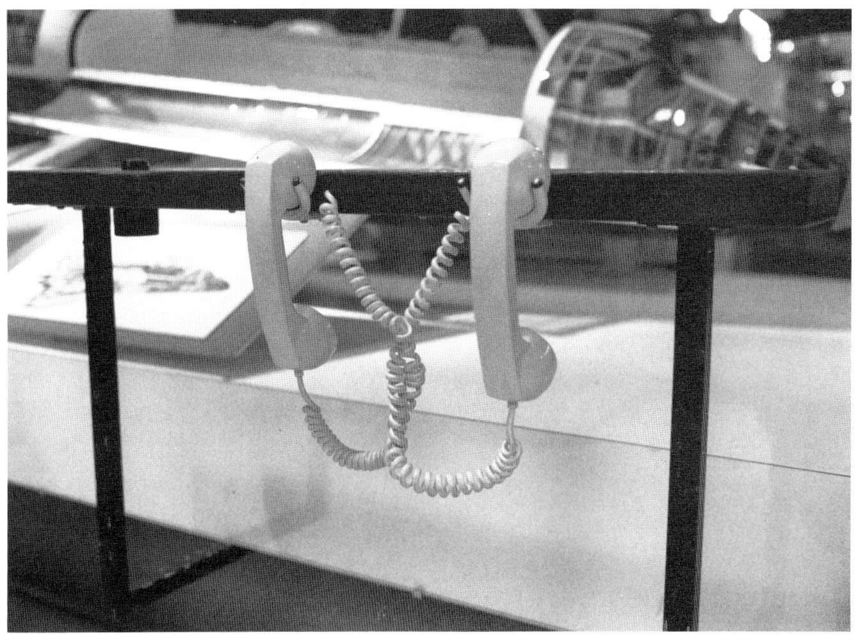

Tangled cords When multiple ear speakers are provided at a station, the cords often become entangled, making them unusable. This is a source of aggravation both for visitors and for the museum maintenance staff. Possible solutions include spacing the phone handles further apart, limiting the number of phones, or making the cords retractable. Each solution presents trade-offs.

The sound room Glass walls make this sound room feel like an integral part of the exhibit and allow people outside to preview the activity inside. The view visitors have out of the glass room minimizes the feeling of confinement and helps them stay oriented.

Sound-isolating rooms

Exhibits where sound is the subject, such as music, speech, or hearing, may require some type of sound-isolating enclosure. Often these are thermopane, glass-walled rooms with doors, that are designed to accommodate six to ten people (or more).

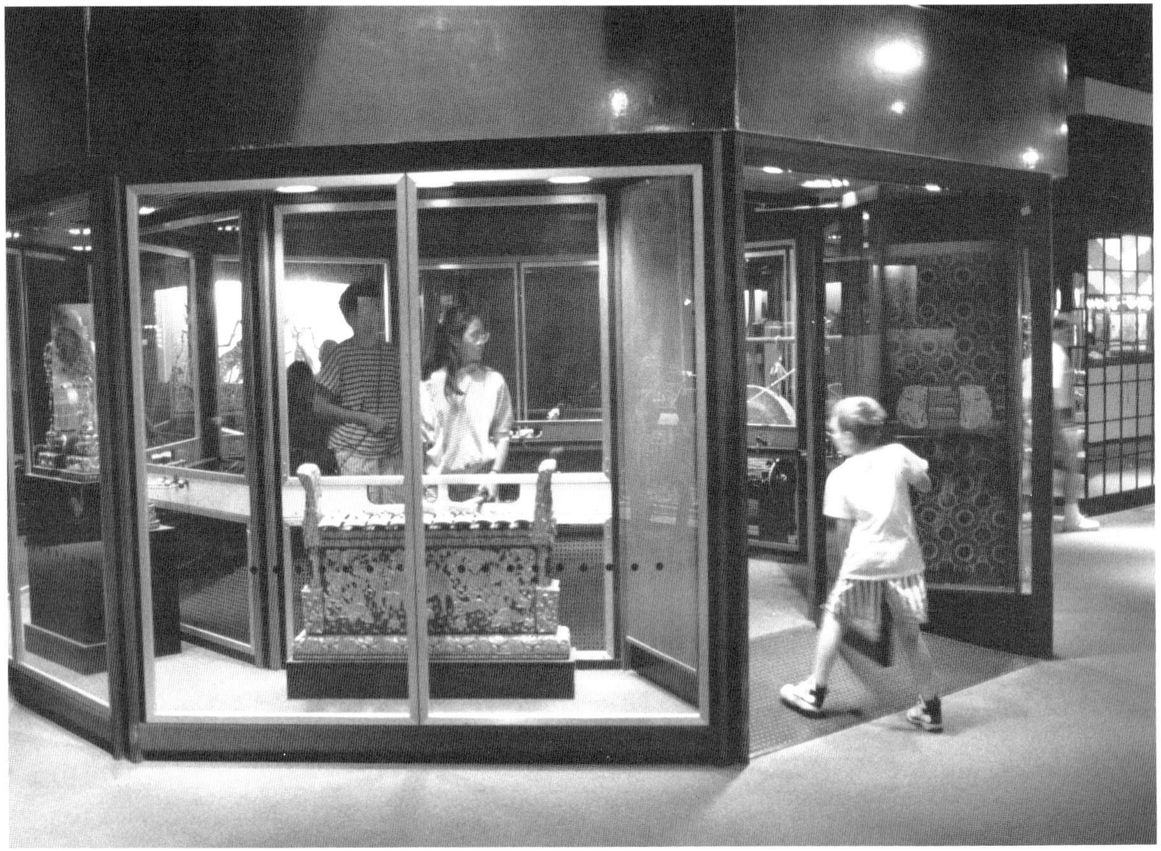

Sound booths

Sound isolation booths designed to accommodate one or two people provide a smaller-scale version of the soundproof room. Commercially manufactured soundproof booths designed for administering hearing tests are quite adaptable to this purpose. In the museum setting a sound-isolating booth serves two functions: isolating the listener from the ambient noise, and containing the sound generated by the exhibit. Glass windows should be specified for sides and doors.

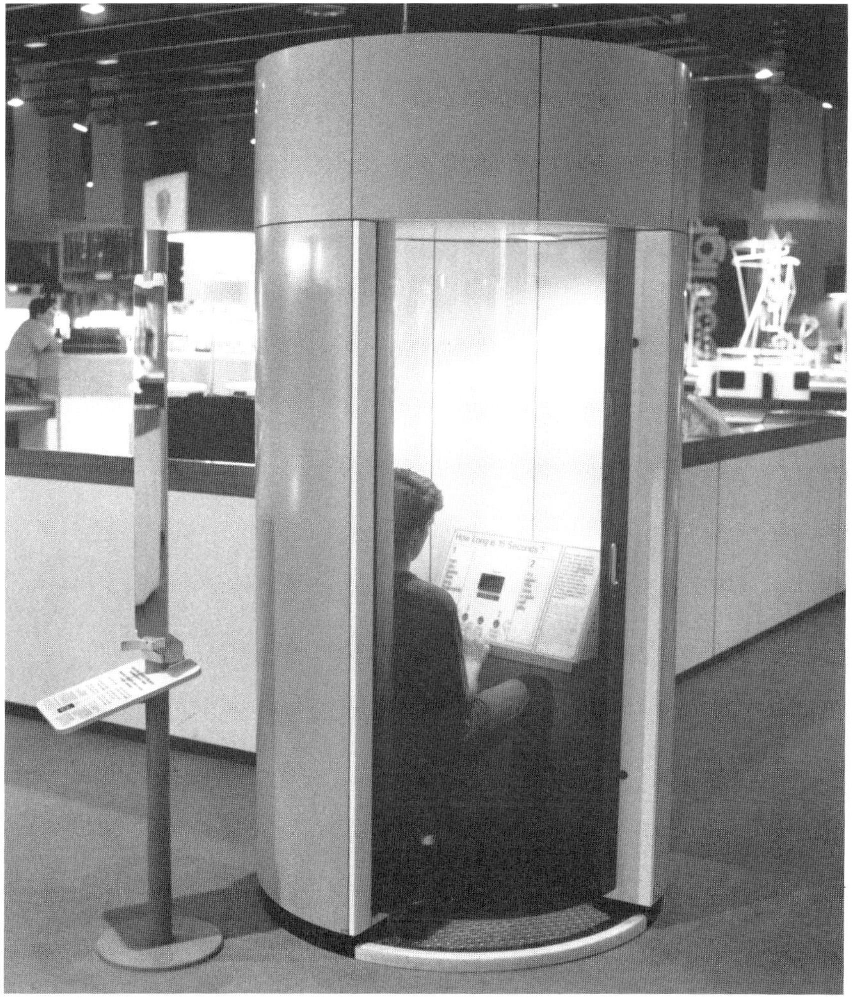

Sound isolation booth
A sound isolating booth is effective both as a sound container and as a means of isolating listeners from distracting ambient noise. The booth shown here is a custom design. The door stays open when the booth is unoccupied to minimize stuffiness and heat build-up.

Water cycle show This small theater presents a multiple-projector slide show and synchronized display lighting. The presentation lasts about 10 minutes. The theater seats 10 people and occupies approximately 250 square feet.

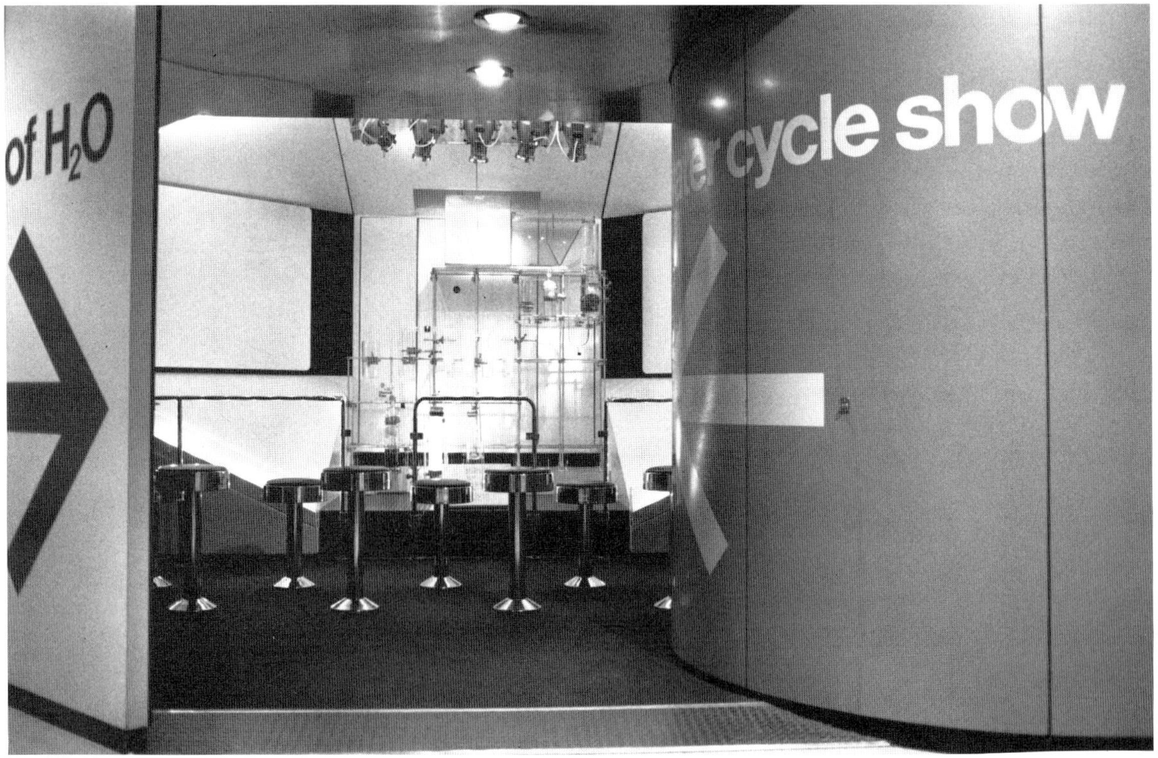

Small theaters

The sound from film and video shows can best be controlled by locating the presentation in a small theater with closable doors, or a sound-dampened, baffled entrance/exit. This approach is widely used in larger exhibits and is typically for shows that engage the audience for eight to ten minutes or longer. Seating for 10 to 40 people is common.

Semi-acoustically isolated exhibit enclosures

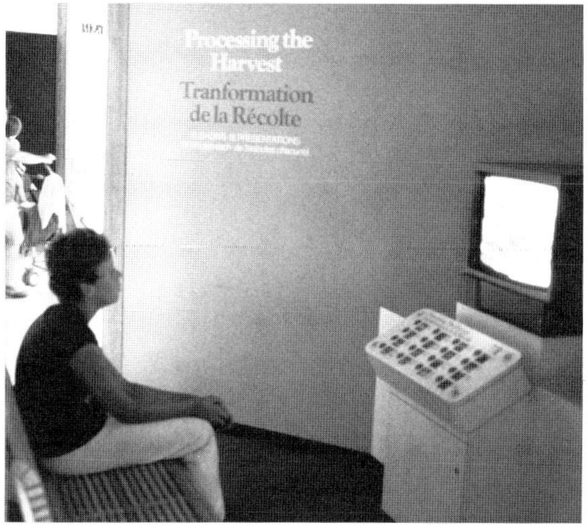

Rest spot In addition to controlling the sound of video presentations, niches provide a place for visitors to take a break from the busy exhibit environment.

Sound-controlled niches

Another way to contain the sound of video presentations is to create a partially enclosed niche. Seating is typically for four to ten people. If presentations are kept short (three to eight minutes) fairly large numbers of visitors can be accommodated. Niches do a good job of containing sound if they are finished with sound-absorbing materials and positioned properly in the overall exhibit space. The quality of the sound presentation can be quite good in these settings.

Semi-enclosed exhibits

Semi-enclosed exhibits are intended for use by one or two people at a time. Typically, enclosure is provided by adding side wings, or a top with speakers enclosed. Sound-absorbing materials on the surfaces of the exhibit inhibit the reflection of sound and absorb ambient sounds. This is a relatively effective way of presenting interactive video exhibits that have audio as an integral feature.

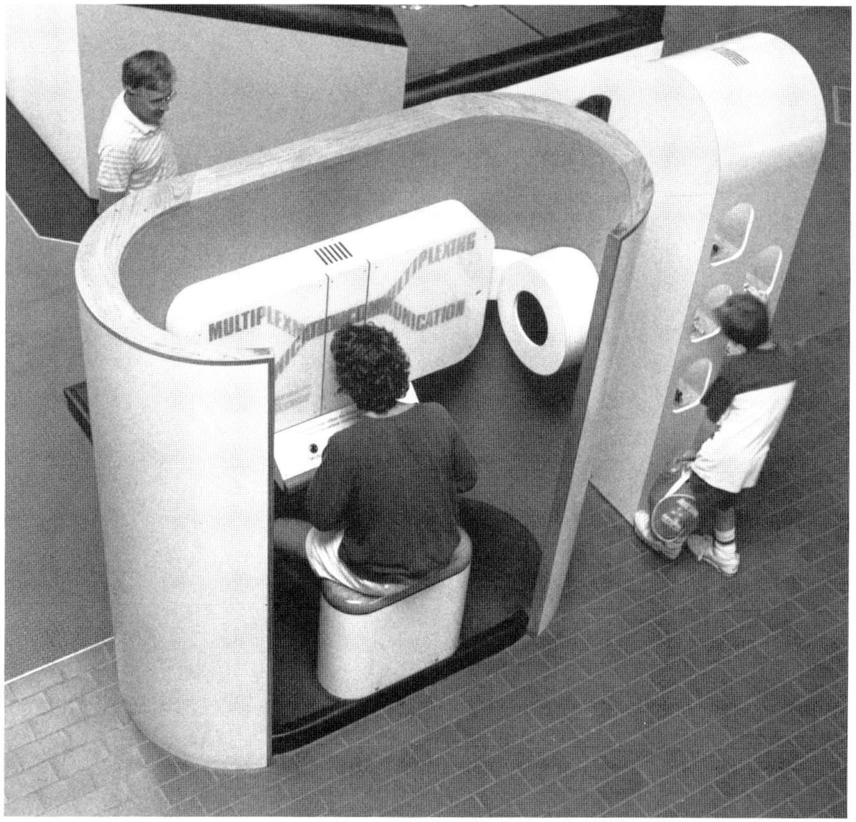

Sound thinking As a general rule, small speakers located close to the listener will minimize sound spill while providing high-quality audio. The two speakers shown here are pointed toward the user's ears. The exhibit's end walls are curved and soft, helping to control the reflection of sound.

Focused sound Parabolic sound reflectors, like the one pictured in the ceiling here, help focus sound. Unfortunately, sound does not stop at a listener's ear, but continues to bounce around the room. As a result, this is a minimally effective means of controlling sound spill.

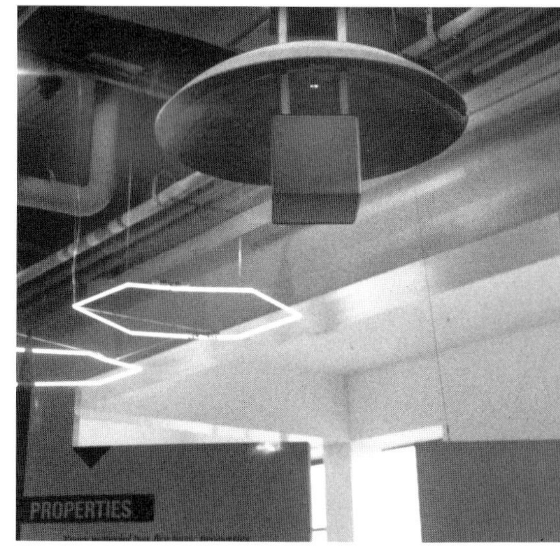

The least desirable approach to audio delivery is to localize sound through speaker placement and volume control, with no form of acoustical enclosure.

This approach sometimes attempts to contain audio by using speakers designed to direct or focus sound. These are minimally effective because once sound waves hit the floor, walls, or other objects, they are re-directed. Also, sound can be heard well only when a visitor is standing directly under a focused speaker.

One of the most significant problems with public audio is that if the volume is loud enough to be heard when the museum is crowded, it is far too loud when the museum is less busy. Fluctuations in levels of ambient sound can be dealt with electronically, using control devices which continuously monitor ambient sound and adjust the volume of a presentation accordingly. These devices add expense to the exhibit, but can greatly reduce noise pollution.

The best way to minimize the noise pollution associated with public audio is by using sound-absorbing materials on ceilings, walls, floors, and large exhibit structures and by dispersing audio presentations widely in the exhibit space.

The benefit of captioning in noisy exhibit spaces should not be overlooked. Video sound tracks that are barely audible due to ambient noise can often be successfully followed when captions appear on the screen.

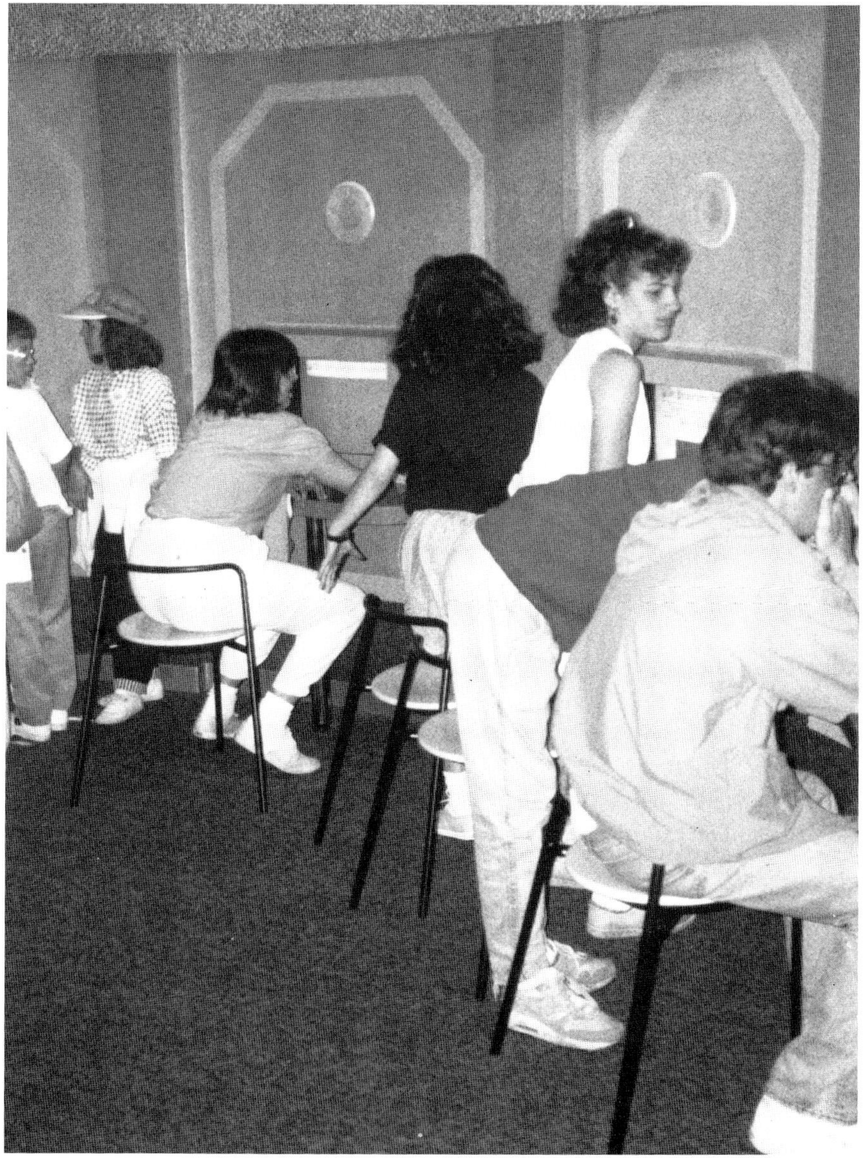

Noise pollution These interactive video stations have no acoustical controls. Speakers aim out into a relatively small exhibit area devoid of sound-absorbing finishes. Audio tracks from adjacent units spill over, creating a cacophony of unintelligible sound. Captions – if they had been provided for hearing impaired visitors – would have also aided those with normal hearing to follow the audio part of the presentation.

7

Exhibit Lighting

The three main aspects of lighting affecting exhibit use are contrast, glare, *and* heat. *These issues apply to exhibits in general, not just to hands-on exhibits. However, since lighting is such an important issue for a visitor's comfort, and one which affects accessibility, a brief discussion is included here.*

Levels of contrast

The primary purpose of exhibit lighting is to provide contrast among elements in the museum. For example, in a large exhibit hall, lighting can be designed to provide a contrast between circulation space and exhibit components. In this way, light provides visual cues which help guide visitors through the exhibit hall. On a smaller scale, lighting of an exhibit component can be designed to focus the viewer's attention on a specific element or elements.

Levels of contrast, while varying, must not be radical. In fact, dramatic shifts in lighting levels, such as going directly from daylight to a dark space, should be avoided. The human eye needs time to adapt to highly contrasting light levels. Radical jumps are uncomfortable and disorienting for all visitors, and for those with sight impairments, the experience can be painful.

The same principle applies to lighting text and graphic panels. Lighting on interpretive panels should not contrast radically with overall lighting. Reading a brightly lit, white text panel in an otherwise darkened environment may be impossible for some visitors, or at best, uncomfortable. In darkened environments, light type on a dark background is preferable. Backlighted text is the most difficult of all to read, especially for people with sight impairments.

ADAPTING TO THEATERS

Small film and video theaters integrated into an exhibit hall present a special light adaptation concern. If the exhibit hall lighting level is bright, people may have a hard time seeing aisles and seating on the way into the darkened theater. At the end of the show, going from the theater back into a brightly lit exhibit hall can also be a jarring experience. Slowly raising and lowering the lighting as part of the theater presentation should be considered as a means of allowing visitors to comfortably adapt their vision to changing light levels. This is particularly helpful for people with sight impairments.

Glare

There are two common glare problems in exhibits. One occurs when lighting is reflected from a plastic or glass case or other highly reflective surface into a viewer's eyes. The other problem occurs when accent lighting is aimed poorly and ends up in a viewer's eyes. By developing a detailed lighting plan during the exhibit design process, it is possible to avoid most glare problems. Consideration should be given to working with an architectural lighting consultant who is familiar with the requirements of exhibits.

The difficulty of controlling glare is compounded when exhibit components are constructed of highly reflective materials such as glossy acrylic plastic or highly lacquered woods. There is really no reason, other than style, to use highly reflective materials for exhibit finishes. When clear plastic is used to protect exhibit labels, thin (1/16 inch) material with a non-glare surface finish should be used to reduce glare.

Lighting and heat

Lighting load of an exhibit and backlighting of large film images are the primary sources of heat gain in an exhibit space. When the air-conditioning system for a major museum construction project is being engineered, the exhibit lighting load is an important factor to be considered. An exhibit space that becomes overheated when the museum is crowded is extremely uncomfortable for visitors.

Labels for Hands-On Exhibits

8

Labels explain how to use hands-on exhibits and provide a wide range of interpretive information. To be effective, labels must be both legible and readable. Effective placement is also an important factor contributing to a visitor's ability to use and understand the exhibit.

Often, a visitor's first response to a hands-on exhibit is to begin playing or exploring. Instructions for operation and interpretive labels are consulted following this period of use. This sequence does not diminish the importance of labels. In fact, a visitor's interest in written information may be stimulated by questions formulated during initial exploration. When the user gets stuck, or is unable to operate the exhibit, instructions become essential reading.

Labels also stimulate interactions among visitors. Written text often serves as a point of departure for a group discussion related to an exhibit activity.[4] A parent with a child reads instructions and interpretive labels while the child explores. It is critical that labels be brief and to the point to help a parent in his or her role of quickly becoming an "explainer."

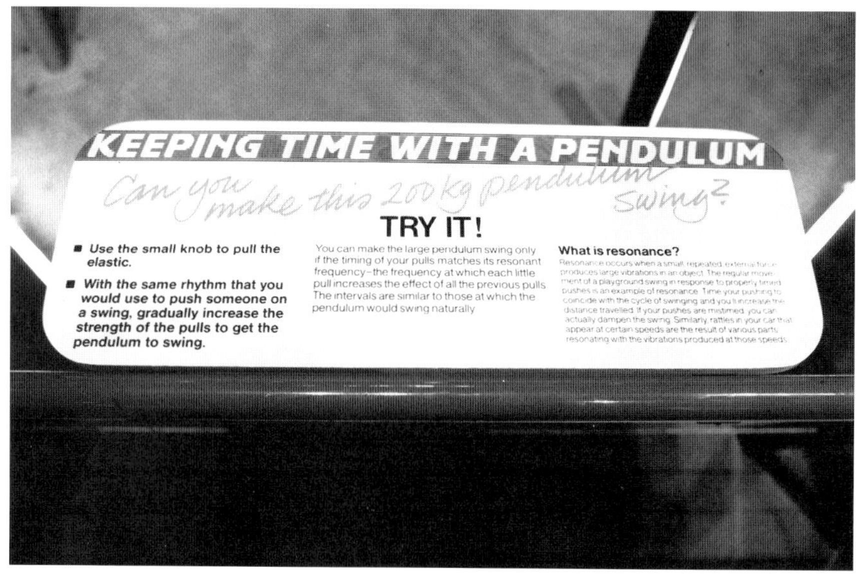

How to use this exhibit The first thing users want to know is how to make the thing in front of them work. Instructions for use typically take the form of a numbered or bulleted list of operations. The list needs to be short, three to four items at most. If more instructions are needed, the design may be too complicated and should be reviewed. The text should be brief, and the language clear.

"Action" labels Visitors can sometimes find helpful cues from "action" labels placed on or next to controls such as *push, pull, turn, lift*, and arrows indicating direction. The handprint in this exhibit provides an additional nonverbal cue.

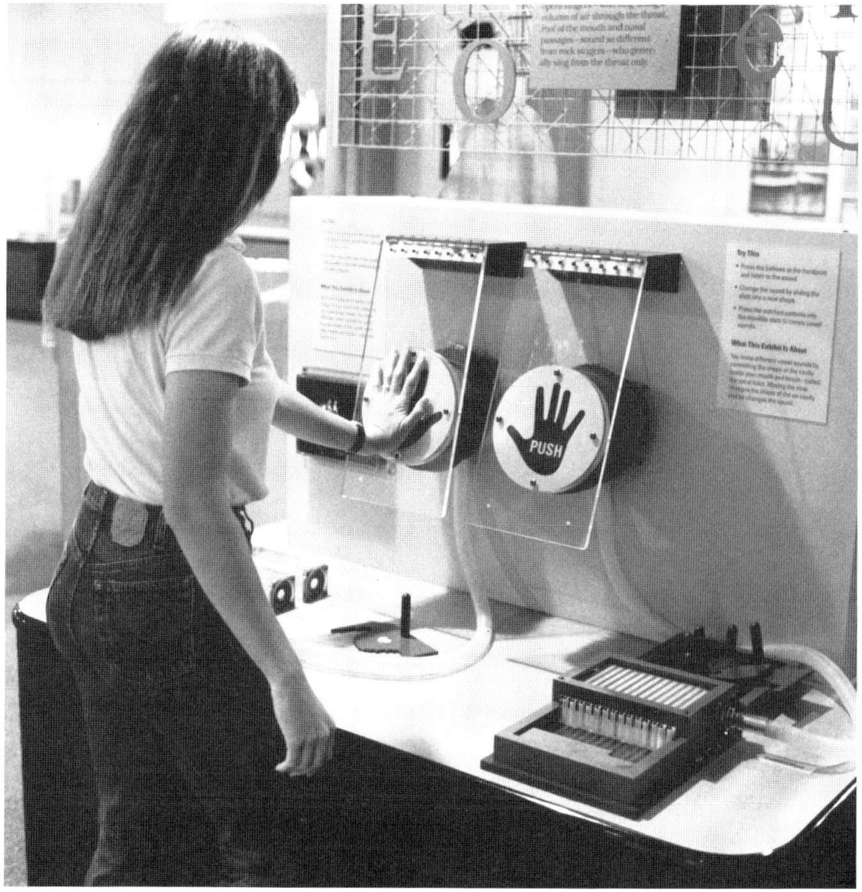

The personal touch Although there are written instructions that provide a fairly clear description of what to do, the possibilities inherent in this open-ended exhibit become evident only when a staff member gives a brief demonstration. Here a demonstrator shows visitors how to fly using a mirror.

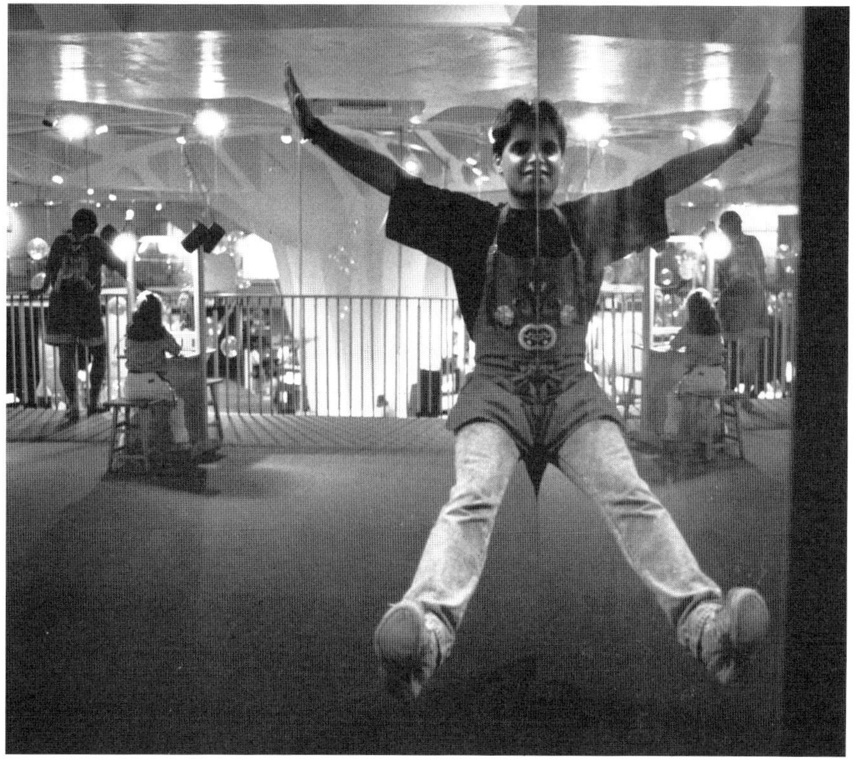

Put it where you use it Instruction labels should be located as close to controls as possible. If not, users may have difficulty finding controls. In this example, the instruction label is around the corner from the control, and an arrow is needed to point users in the right direction.

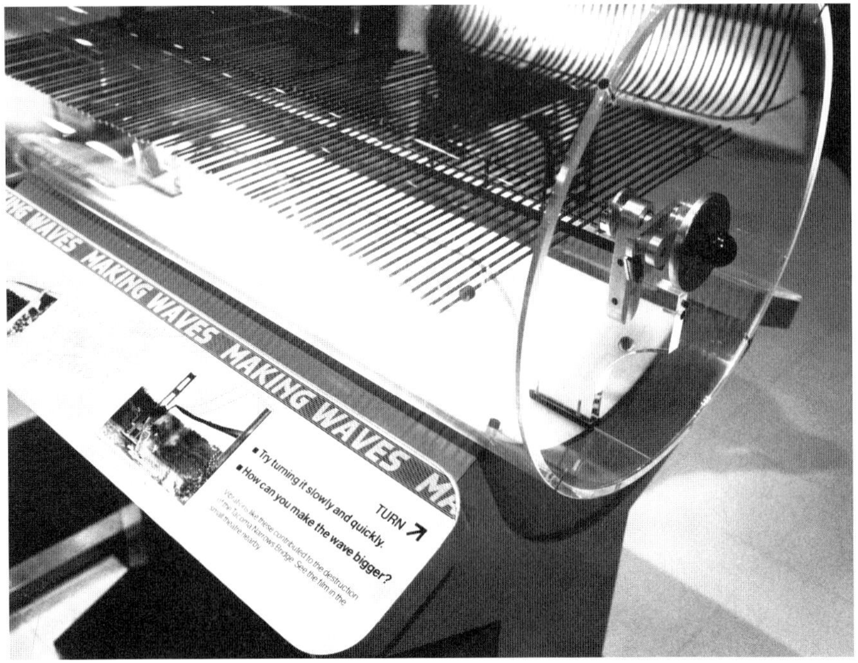

Hidden labels These labels, located along the back edge of the casework, are obstructed from view by artifacts and hands-on exhibits. A scale model or full-size mock-up would have averted this problem.

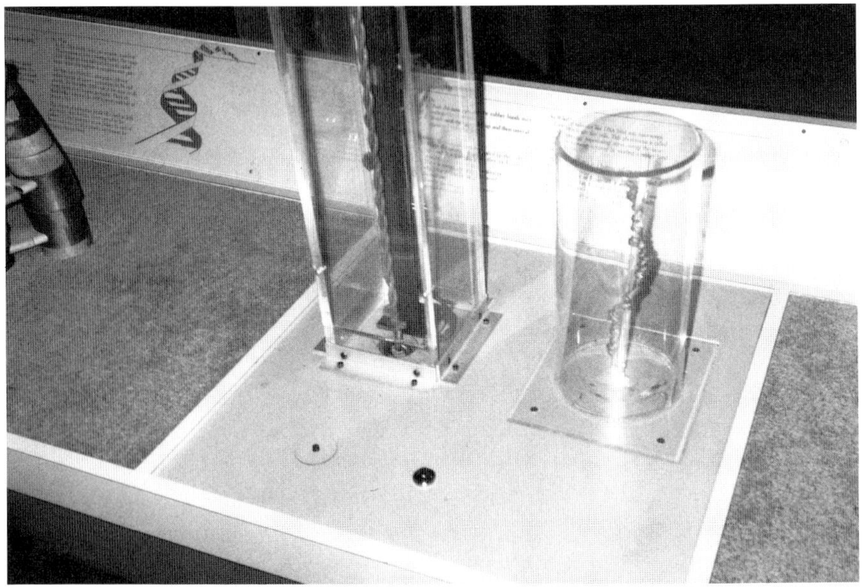

A picture is worth a thousand words This exhibit is not particularly complex, but its use is difficult to explain in words. This problem is overcome with a drawing that shows someone using the exhibit. Drawings or photos showing exhibits in use are also a good way of overcoming language barriers.

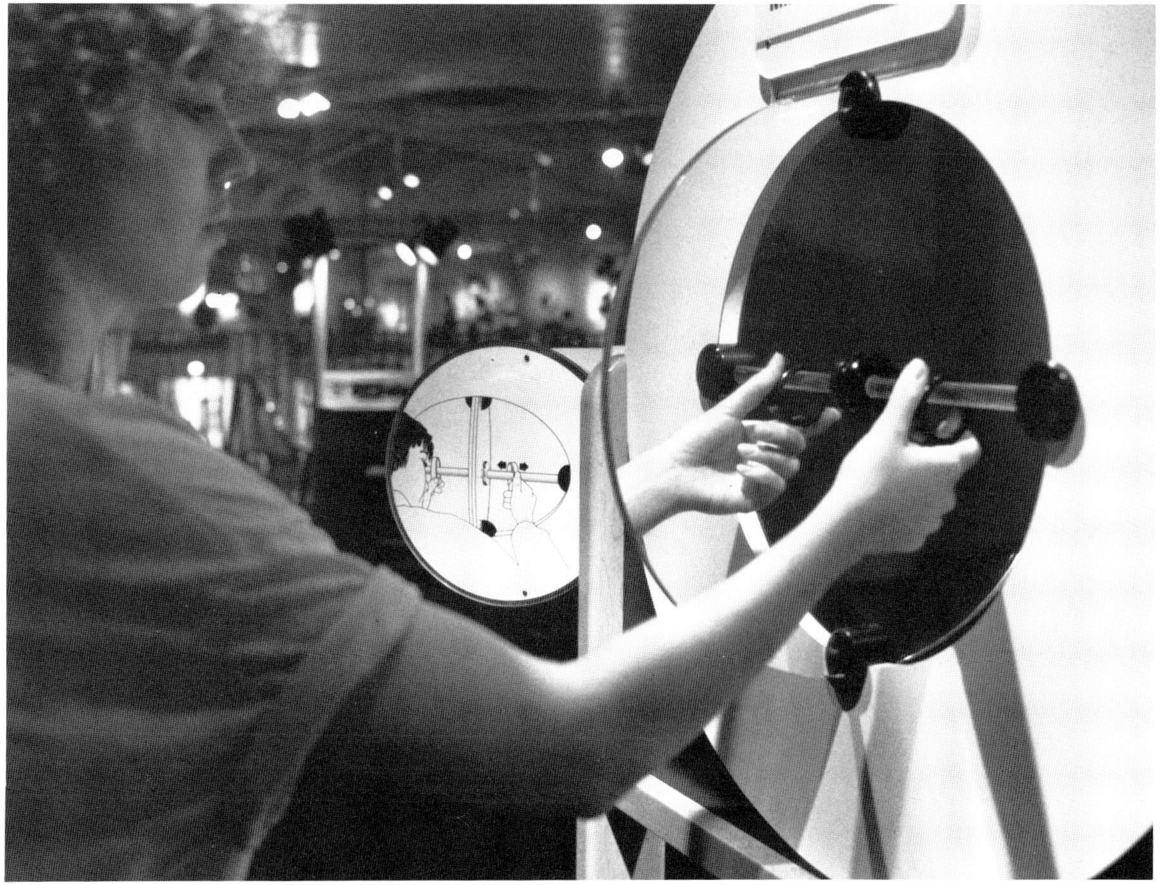

GUIDELINES FOR LABEL PLACEMENT

1. Place labels near the part of the exhibit being interpreted.
2. Position labels so that individuals with low vision can get near for close-up reading.
3. Angle labels or place them low enough so that individuals in wheelchairs and those wearing bifocals can read comfortably.

Legibility

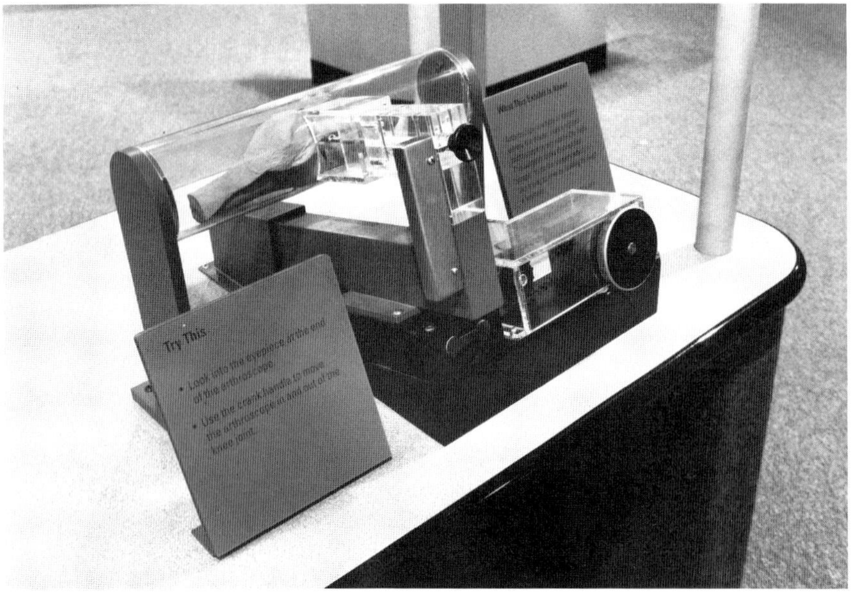

Changing labels
Exhibit instructions and interpretive labels should be prototyped and evaluated along with the rest of the exhibit. When the exhibit is first opened to visitors, the need for additional label revisions often becomes apparent. Labels should be designed to be easily removed and inexpensively replaced. The ones shown are laser printed on colored paper and covered with thin, nonglare plastic.

Exhibit labels should be able to be read comfortably by a wide range of visitors including older people, many of whom experience some degree of sight impairment. All too often, though, labels are too small, have too little contrast between type and background, or are set with type that is too loose or too tight. This makes reading uncomfortable and tiring for everyone.

The most common reason for poor legibility is poor use of type. Good typographic design involves more than selecting a legible typeface and setting it at a standard size. In fact, legibility is more affected by line length, letter spacing, word spacing, and line spacing than by the style or size of the typeface. Here are some general guidelines to keep in mind when developing labels intended for use by a wide range of visitors.

Color and contrast

The most legible text combines black type on a white background. But selecting color combinations for type and background is not that simple. For example, exhibit lighting also is a factor in selecting type and background combinations.

If the exhibit has a very low ambient lighting level, then bright white backgrounds cause glare and create an eye adjustment problem. For people with sight impairments, the glare can be a painful distraction. In darker environments, a dark background with light type is preferable. Whether choosing light or dark, it is critical to provide a high level of contrast between the type and the background. Subtlety is very bad for legibility.

Typography

The following rules of thumb apply to labels intended to be read from a distance of three feet or less. Guidelines are good for starters, but there is no substitute for full-size label mock-ups evaluated in conjunction with exhibit prototypes. It also helps to work with a designer strong in typographic skills.

- *Typeface:* The following typefaces are very legible when set correctly: Times Roman, Palatino, Century, Helvetica, and Universe.

- *Type weight:* Text should be set in the medium or book weight of the chosen typeface. Headlines should be set in a contrasting weight or typeface to set them apart from the text.

- *Type size:* Text should be set in 18- to 30-point type, depending on the typeface chosen. Headlines should be set in the same size or larger. A limited field of vision makes it difficult for an individual with low vision to read information set in extremely large sizes.

- *Type styles:* Type set in all caps is much more difficult to read than type set in upper and lower case. Never use all caps for text. Large amounts of text set in italic are also more difficult to read than text set in Roman type and reading slows as a result. Use italics only for emphasis in the text.

- *Letter and word spacing:* Correct spacing between words and letters creates an even appearance which enhances the flow of reading. If letter spacing is too tight, it becomes more difficult to see the shapes of the letters and words. If word spacing is too loose, white spaces between the words interrupt the flow of reading.

- *Line length:* Line lengths for text should average 45 to 50 characters per line. Because of the way the eye scans a block of text, lines of type with too many characters are difficult to follow and those with too few characters are more tiring to read.

- *Line spacing:* Spaces between the lines of type allow the eye to scan across the page and then find the next line easily. Determing spacing depends on the size of the type, the length of the lines, and the particular typeface being used. Too little spacing between lines makes text very difficult to read.

• *Margins:* Flush left, ragged right margins are easiest to read. The uneven lines on the right provide a visual clue about where the reader is in the text block. Also, consistency of word and letter spacing is easily maintained throughout the text.

• *Hyphenation:* It should to be kept to a minimum. Try to avoid having two hyphens in a row, and watch for situations where hyphenation makes it difficult to follow the sense of the text.

Creating labels that work for the audience requires a team effort including the program developer, exhibit designer, typographic designer, and label writer. Together the team can synthesize content and exhibit objectives and create interpretive, editorial, and design guidelines for labels and graphics. Following these guidelines will help to create labels with a look, feel, and flavor personalized for a particular institution.

Based on our case studies, people *do* read hands-on exhibit labels. These labels need to be both legible and readable. It helps to keep in mind the following:

• *Use of language:* Museum labels need to be clear, to the point, and free of jargon.

• *Quick access:* Headlines and subheads can be written and designed to provide quick access to label content. Visitors browse through headings to zero in on items of interest.

• *Consistency:* If interpretive graphics are developed as a system that is used consistently, visitors can learn to find the level and kind of information they are most interested in.

• *Hierarchy:* It is helpful to provide readers with visual clues about what might be most important to read. Clues can be provided by the size of type (bigger is often more important) and label location (a label located next to a hands-on component may be more important than a nearby label in larger type).

There are as many kinds of interpretive labels as there are exhibits. The following examples show two very different approaches. Each reflects the exhibit topic, the institution's philosophy, and its audience.

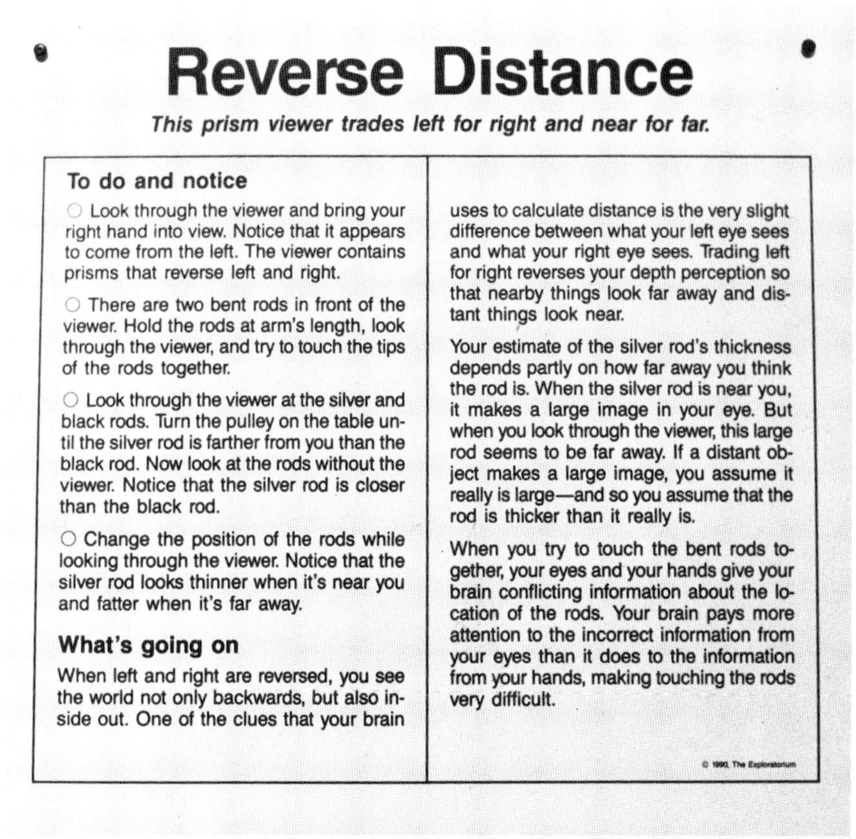

Reverse Distance

This prism viewer trades left for right and near for far.

To do and notice

○ Look through the viewer and bring your right hand into view. Notice that it appears to come from the left. The viewer contains prisms that reverse left and right.

○ There are two bent rods in front of the viewer. Hold the rods at arm's length, look through the viewer, and try to touch the tips of the rods together.

○ Look through the viewer at the silver and black rods. Turn the pulley on the table until the silver rod is farther from you than the black rod. Now look at the rods without the viewer. Notice that the silver rod is closer than the black rod.

○ Change the position of the rods while looking through the viewer. Notice that the silver rod looks thinner when it's near you and fatter when it's far away.

What's going on

When left and right are reversed, you see the world not only backwards, but also inside out. One of the clues that your brain uses to calculate distance is the very slight difference between what your left eye sees and what your right eye sees. Trading left for right reverses your depth perception so that nearby things look far away and distant things look near.

Your estimate of the silver rod's thickness depends partly on how far away you think the rod is. When the silver rod is near you, it makes a large image in your eye. But when you look through the viewer, this large rod seems to be far away. If a distant object makes a large image, you assume it really is large—and so you assume that the rod is thicker than it really is.

When you try to touch the bent rods together, your eyes and your hands give your brain conflicting information about the location of the rods. Your brain pays more attention to the incorrect information from your eyes than it does to the information from your hands, making touching the rods very difficult.

© 1990, The Exploratorium

Two-topic label
This example shows an approach to labels developed at the Exploratorium. Subheads are used to make a clear distinction between operating the exhibit ("To do and notice") and interpreting the phenomenon ("What is going on?"). This reflects a typical user pattern of experimenting with the apparatus first, and then reading the interpretive text. These labels are long and detailed. This reflects the goals of the exhibit and the philosophy of the institution.

Nonverbal interpretation One of the goals of this exhibit on electricity and magnetism is to communicate as much as possible nonverbally. Instructions and interpretation both rely heavily on a variety of pictorial labels. This approach reflects the California Museum of Science and Industry's desire to make exhibits accessible to those visitors who do not read English.

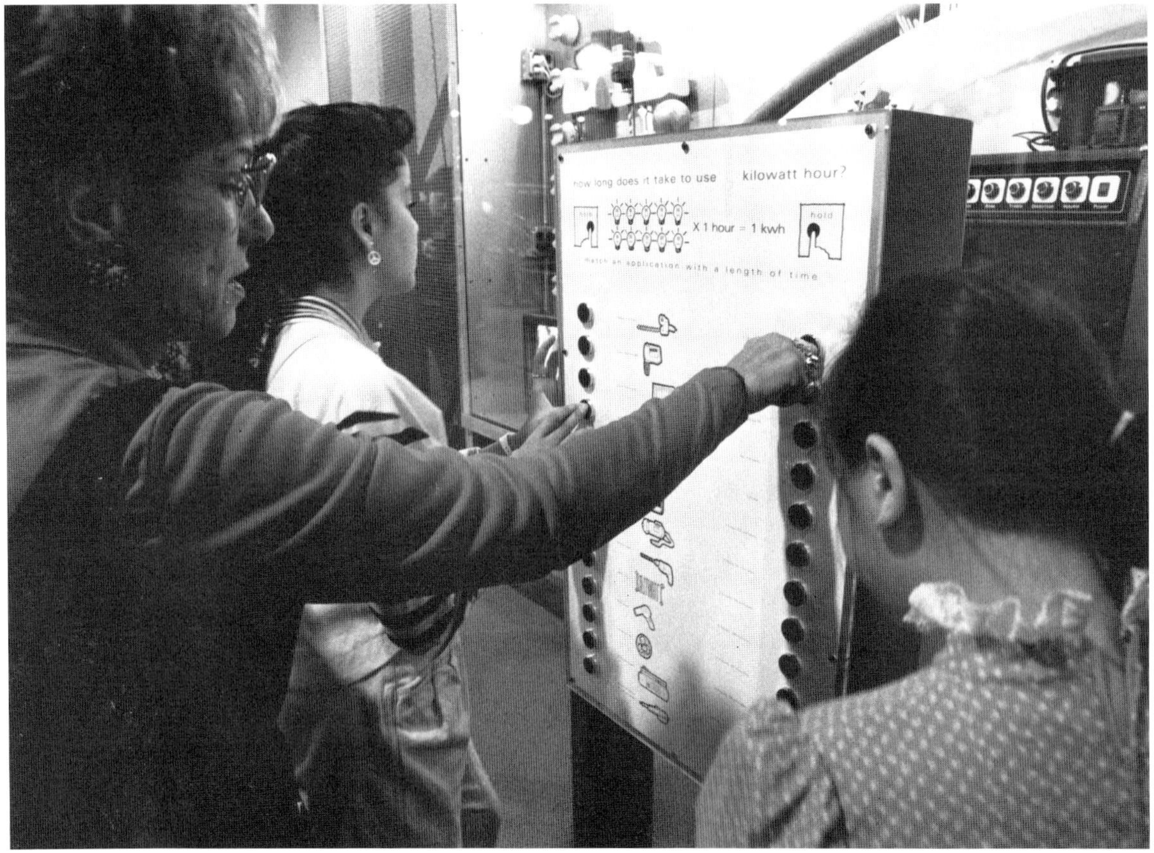

LABEL GUIDELINES FOR PEOPLE WHO ARE DEAF OR HAVE SEVERE HEARING IMPAIRMENTS

People who become deaf or develop severe hearing impairments before learning to talk rarely develop a full range of language skills later in life. Labels accessible to these visitors should follow the guidelines below. Following these guidelines will result in labels that are clearer and more comprehensible for all.

1. Let each sentence express one thought.
2. Keep sentence structure simple.
3. Avoid embedded phrases.
4. Use everyday vocabulary.
5. Define unfamiliar words by using parentheses or the word "or" in the text.

User Friendly Exhibit Design Checklist

9

Designing a hands-on exhibit that is "user friendly" means bearing in mind literally dozens of issues – consulting with experts in many fields, prototyping and testing extensively, and remembering that the visitor's experience is of paramount importance. This checklist should serve as a point of reference for exhibit development team members throughout the design process. Maintenance and safety are included here because they are a major concern when designing hands-on exhibits. Finally, even a comprehensive checklist is no substitute for prototyping and testing – when in doubt, try it out.

General Issues

- ❑ Exhibits are designed for an audience diverse in age, size, learning style, level of knowledge, cultural background, and physical ability.
- ❑ Both individual components and the overall exhibit space are designed to minimize ambient noise.
- ❑ Exhibits are usable and understandable on their own. They do not frustrate, inhibit, confuse, or frighten users.
- ❑ How to operate exhibits is communicated primarily through nonverbal clues.
- ❑ Interaction is not forced on concepts that do not lend themselves to this medium.
- ❑ Prototyping and evaluation are integral to the design process.

Human Measurement

- ❑ Exhibits are configured to accommodate the varied physical requirements of a range of visitors.
- ❑ Exhibits are accessible to people using wheelchairs and walking aids.
- ❑ Alternatives to the "one-size-fits-all" approach are considered during the exhibit design process.
- ❑ Mocks-ups are used to evaluate exhibit dimensioning, including the relative location of the user, controls, labels, and exhibit apparatus.

Controls

- ❑ All control apparatus directly related to the exhibit concept is visible to users.
- ❑ Where possible, controls are designed to reinforce exhibit concepts.
- ❑ Controls are selected to communicate visually how they are used: round handles are for turning, levers are for pulling, buttons are for pushing.
- ❑ Control labels are used where necessary to tell users what to do: "turn," "pull," "push."
- ❑ As few controls as possible are required to operate exhibits.
- ❑ Controls for different functions look different.
- ❑ Controls are located within 10 inches of the front of tabletop exhibits, providing access for people with motor control impairments.

- ❑ Where multiple controls are required, the number of controls is matched to the number of functions, and the controls are organized according to function.
- ❑ Feedback is incorporated into the control mechanism.
- ❑ Mechanical controls are matched to performance requirements. Type of motion, range of motion required, degree of precision required, range of operating speed, and user power capability are all considered.
- ❑ Controls are designed for safe operation. Handles that spin rapidly do not have spokes. Clutches are employed to separate controls from the mechanical energy of exhibit apparatus.
- ❑ Controls are designed for long-term durability.
- ❑ Controls are evaluated during exhibit prototyping. The "feel" of controls is fine-tuned during final fabrication.

Eye Pieces and Viewers

- ❑ Viewing devices are designed for comfortable and safe use.
- ❑ Hand-held viewers are used as an alternative to fixed viewers wherever possible.
- ❑ Alternatives to laboratory-style microscopes have been considered.

Interactive Video and Computers

- ❑ Captioning is provided on all video presentations.
- ❑ Multiple interactive stations of the same program are provided where space and funds allow.
- ❑ Groupings of stations that include audio are designed and located to minimize sound spill between stations.
- ❑ Seating at single-user stations can accommodate two people.
- ❑ Seating is provided where extended exhibit use is anticipated or desired.
- ❑ Interactive video and computer monitors are frontally accessible to wheelchair users and are mounted at an easily reachable height.
- ❑ Input devices for interactive computers and videodiscs are accessible to wheelchair users.
- ❑ Input devices have been selected based on criteria for specific user tasks as well as general criteria such as cost, reliability, and maintenance.

Interactive Video and Computers

- ❑ Complex interactive programs have been planned and budgeted to allow adequate time and funding for formative evaluation, advance prototyping and evaluation, and summative evaluation.

Audio Exhibits

- ❑ Audio delivery systems have been selected or designed to minimize noise pollution.
- ❑ The exhibit environment uses sound-absorbing materials on the walls, floor, and ceiling to help control sound.
- ❑ Small, enclosed rooms designed to contain sound have large window areas to connect people inside with their surroundings and to show people outside the activity within.
- ❑ Off-the-shelf headphones are beefed up to withstand the rigors of the hands-on exhibit environment.
- ❑ Audiovisual exhibits three minutes or longer are designed to function as rest stations as well as viewing and listening areas.

Lighting

- ❑ Lighting is used to help guide people through the exhibit.
- ❑ Lighting levels change gradually, allowing for adaptation and minimizing the discomfort of people whose eyes are particularly light sensitive.
- ❑ Lighting does not bounce off reflective surfaces into visitors' eyes, regardless of their viewing height or angle.
- ❑ Fixtures are aimed so that people do not have to look into bright lights, regardless of their viewpoint.
- ❑ Lighting has been considered in the heat load of the exhibit area.

Labels

- ❑ Labels are clear, concise, and to the point.
- ❑ Operating instructions are located next to controls.
- ❑ When written words cannot explain how to use exhibits clearly, illustrations, diagrams, or photos are used.
- ❑ Labels are designed to be replaced easily and inexpensively.
- ❑ Typeface, typographic design, layout and colors, and exhibit lighting all work to promote legibility of labels.
- ❑ Backlighted text is used only as a last resort.
- ❑ Labels are drafted and evaluated during the prototyping phase of exhibit design.

Maintenance

- ❑ The museum's maintenance capabilities, including staffing and facilities, have been assessed before beginning exhibit design.
- ❑ Exhibits do not exceed the maintenance capabilities of the museum.
- ❑ Exhibits are designed to be as rugged and low maintenance as possible.
- ❑ Exhibits are outfitted with the most reliable, most durable components available. Although this will increase the front-end cost, reduced maintenance will save money, precious time, and aggravation in the long run.
- ❑ Exhibit design provides convenient maintenance features such as easy access into exhibit housing; adequate room in housing for accessing components for adjustment, repair, or removal; and quick-access panels to get at regularly replaced items.
- ❑ Exhibits are designed so that individual components can be removed for maintenance without leaving an obvious, gaping hole.
- ❑ Hardware is standardized throughout the exhibit and is readily available from local sources.
- ❑ Parts inside exhibit housing to be adjusted by staff are clearly labeled.
- ❑ Exhibits include the following documentation: as-built plans; electronic schematics; an itemized list of parts and sources for regularly replaced items; an operating manual with a trouble-shooting guide and maintenance requirements; and a master schedule for keeping track of routine maintenance such as bulb changing, audio tape replacement, and lubrication.
- ❑ Maintenance issues have been discussed during prototyping.

- ❑ Exhibit structures are stable even when visitors lean and sit on them in ways not intended by designers.
- ❑ Protective guards or covers are placed over moving parts that could be a source of injury.
- ❑ All components have been checked against "finger pinching," one of the most common museum injuries.
- ❑ When glass is used in exhibits, it is safety laminated.
- ❑ Acrylic plastic is used as a glass substitute wherever possible. Polycarbonate plastic such as Lexan is used where extra resistance to breakage is required.
- ❑ All electrical switching controlled by visitors is low voltage.
- ❑ Exhibits using both water and electricity use ground fault interrupt (GFI) breakers.
- ❑ All high voltages inside exhibit structures are clearly labeled.
- ❑ Exhibits that present potential heat build-up problems are well ventilated and have redundant safety shut-offs when a predetermined peak temperature is reached.
- ❑ Handles that spin do not have spokes.
- ❑ All controls that transfer energy from users to exhibits are designed so that energy from the exhibit cannot be transferred back to the user.
- ❑ Exhibit structures at head level do not extend beyond the perimeter of the base.
- ❑ Edges of exhibit structures, especially those at the head height of young children, are not sharp or pointed.
- ❑ Exhibit designs have been carefully analyzed for potential safety hazards, and prototypes have been used to discover unanticipated safety problems.

1. Norman, Donald A. *The Design of Everyday Things,* pages 13-33.

2. Diffrient, N., Tilley, A.R., and Bardagiy, J. *Humanscale 1/2/3.*

   Diffrient, N., Tilley, A.R., and Harman, D. *Humanscale 4/5/6.*

   Diffrient, N., Tilley, A.R., and Harman, D. *Humanscale 7/8/9.*

3. Flagg, Barbara, "Summary of Implementation Formative Evaluation of *Earth Over Time*" Videodisc, page 42.

4. McManus, Paulette M. "Watch Your Language! People Do Read Labels" *What Research Says about Learning in Science Museums*, pages 4-6.